(P)
chnology
oday Series
Series for Technicians

Mathematics for Techn

Scotec Edition

Level III Engineering Mathematics

Mathematics for Technicians

Scotec Edition

Level III Engineering Mathematics

R. B. Buchan

B.Sc., F.S.S., A.F.I.M.A.,
Head of Mathematics
Esk Valley College

A. Greer

C.Eng., M.R.Ae.S.
Senior Lecturer
City of Gloucester College of Technology

G. W. Taylor

B.Sc.(Eng.), C.Eng., M.I.Mech.E.
Principal Lecturer
City of Gloucester College of Technology

Stanley Thornes (Publishers) Ltd.

First published in 1981 by:
Stanley Thornes (Publishers) Ltd
EDUCA House
Liddington Estate
Leckhampton Road
CHELTENHAM GL53 0DN
England

British Library Cataloguing in Publication Data
Mathematics for engineering technicians.
Level 3: Engineering mathematics for Scotec
1. Shop mathematics
I. Buchan, Robert B
II. Greer, Alec
III. Taylor, Graham William
510'.2'46 TJ1165

ISBN 0–85950–443–3

Typeset in Monotype 10/12 Times New Roman by
Gloucester Typesetting Co Ltd
Printed and bound in Great Britain by
The Pitman Press, Bath

CONTENTS

AUTHORS' NOTE ON THE SERIES

Arising from the recommendation of the Hudson Report, with particular reference to the Haslegrave Report, the Scottish Technical Education Council was set up in 1973. The Council have established six Sector Committees each responsible for a group of courses.

A major change of emphasis in the educational approach adopted in each SCOTEC course has been introduced by the use of 'objectives' in the syllabus which allows student and lecturer to achieve planned progress through the course on a step-by-step basis.

This set of books started with *An Introductory Course* which provides the mathematics required by Level I Technicians and continues at Level II with sister volumes, *Electrical Engineering Mathematics* and *Mechanical Engineering Mathematics*, and concludes with this volume at Level III, *Engineering Mathematics*. Each book follows a standard pattern, and each chapter opens with the words "After reaching the end of this chapter you should be able to:-" and this statement is followed by the objectives for that particular topic as laid down in the Syllabus. Thereafter each chapter contains explanatory text, worked examples, and copious supplies of further exercises. As planned at present the series comprises:-

AN INTRODUCTORY COURSE	Level I
MECHANICAL ENGINEERING MATHEMATICS	Level II
ELECTRICAL ENGINEERING MATHEMATICS	Level II
MATHEMATICS FOR ENGINEERING TECHNICIANS	Level III

R. B. Buchan
A. Greer
G. W. Taylor

1981

1. GRADIENT OF A CURVE

After reaching the end of this chapter you should be able to:-

1. *Determine gradients of chord and tangent to a simple curve.*
2. *Deduce that the process of moving a point on a curve towards a fixed point on the curve causes the gradient of the chord joining the points to approach that of the tangent to the curve at the fixed points.*
3. *Identify incremental changes in the* x, y *directions as* δx *and* δy.
4. *Determine the ratio* $\frac{\delta y}{\delta x}$ *in terms of* x *and* δx *for the function in (3).*
5. *Demonstrate graphically that the rate of change of* $\sin\theta$ *is* $\cos\theta$.
6. *Identify a function of a function and determines its differential coefficient.*

GRAPHICAL METHOD

In mathematics and engineering we often need to know the rate of change of one variable with respect to another. For instance, speed is the rate of change of distance with respect to time, and acceleration is the rate of change of velocity with respect to time.

Consider the graph of $y = x^2$, part of which is shown in Fig. 1.1. As the values of x increase so do the values of y, but they do not increase at the same rate. A glance at the portion of the curve shown shows that the values of y increase faster when x is large, because the gradient of the curve is increasing.

To find the rate of change of y with respect to x at a particular point we need to find the gradient of the curve at that point.

If we draw a tangent to the curve at the point the gradient of the tangent will be the same as the gradient of the curve.

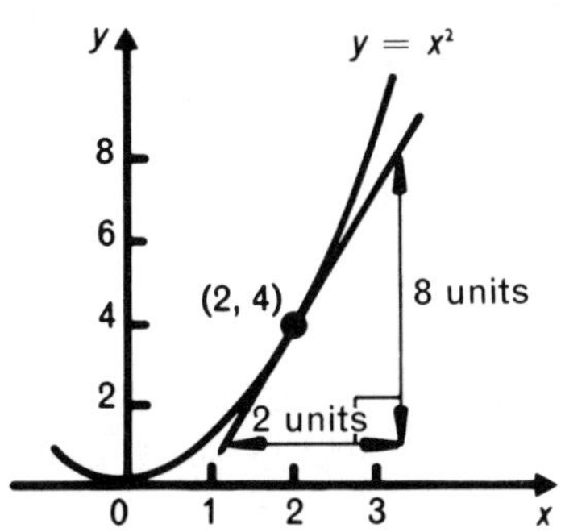

Fig. 1.1

EXAMPLE 1

Find the gradient of the curve $y = x^2$ at the point where $x = 2$.

The point where $x = 2$ is the point (2, 4). We draw a tangent at this point, as shown in Fig. 1.1. Then by constructing a right-angled triangle the gradient is found to be $\frac{8}{2} = 4$. This gradient is positive, in accordance with our previous work, since the tangent slopes upwards from left to right.

EXAMPLE 2

Find the gradient of the curve $y = x^2$ at the point where $x = -2$.

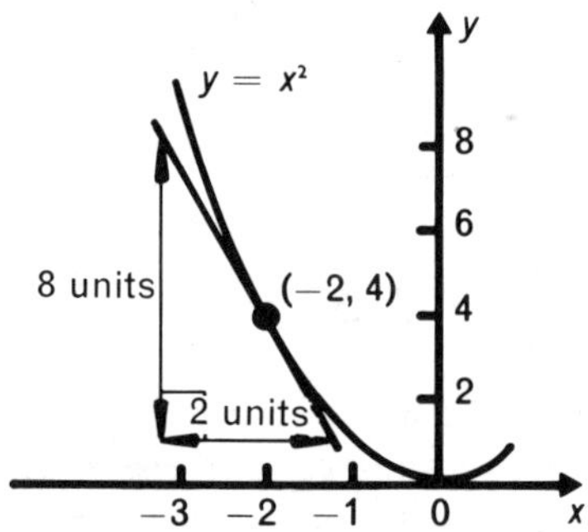

Fig. 1.2

The point where $x = -2$ is the point $(-2, 4)$. By drawing a tangent here, as in Fig. 1.2, and constructing a right-angled triangle as before, the gradient is found to be $\frac{8}{-2} = -4$. This gradient is negative since the tangent slopes downwards from left to right.

EXAMPLE 3

Draw the graph of $y = x^2 - 3x + 7$ between $x = -4$ and $x = 3$ and hence find the gradient at:
(a) the point $x = -3$,
(b) the point $x = 2$.

(a) At the point where $x = -3$,

$$y = (-3)^2 - 3(-3) + 7 = 25$$

At the point $(-3, 25)$ draw a tangent as shown in Fig. 1.3. The gradient is found by drawing a right-angled triangle (which should be as large as conveniently possible for accuracy) as shown, and measuring its height and base.

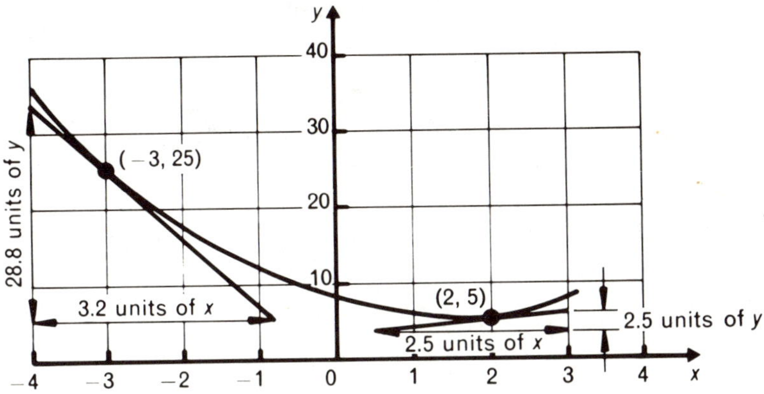

Fig. 1.3

Hence

$$\text{gradient at point } (-3, 25) = -\frac{28.8}{3.2} = -9$$

the negative sign indicating a downward slope from left to right.

(b) At the point where $x = 2$,

$$y = 2^2 - 3(2) + 7 = 5$$

Hence by drawing a tangent and a right-angled triangle at the point (2, 5) in a similar manner to above,

$$\text{gradient at point } (2, 5) = \frac{2.5}{2.5} = 1$$

being positive as the tangent slopes upwards from left to right.

Exercise 1

1) Find the gradient of the curve $y = 3x^2+7x+3$ at the points where $x = -2$ and $x = 2$.

2) Find the gradient of the curve $y = 2x^3-7x^2+5x-3$ at the points where $x = -1.5$, $x = 0$ and $x = 3$.

3) Draw the graph of $2x^2-5$ for values of x between -2 and $+3$. Draw, as accurately as possible, the tangents to the curve at the points where $x = -1$ and $x = +2$ and hence find the gradient of the curve at these points.

4) Draw the curve $y = x^2-3x+2$ from $x = 2.5$ to $x = 3.5$ and find its gradient at the point where $x = 3$.

5) For what values of x is the gradient of the curve

$$y = \frac{x^3}{3}+\frac{x^2}{2}-33x+7$$

equal to 3?

6) For what values of x is the gradient of the curve $y = 3+4x-x^2$ equal to:

(a) -1, (b) 0, (c) 2?

7) Draw the curve

$$y = x-\frac{1}{x}$$

from $x = 0.8$ to 1.2. Find its gradient at $x = 1$.

NUMERICAL METHOD

The gradient of a curve may always be found by graphical means but this method is often inconvenient. A numerical method will now be developed.

Consider the curve $y = x^2$. Part of this is shown in Fig. 1.4. Let P be the point on the curve at which $x = 1$ and $y = 1$. Q is a variable point on the curve, which will be considered to start at the point (2, 4) and move down the curve towards P, rather like a bead slides down a wire.

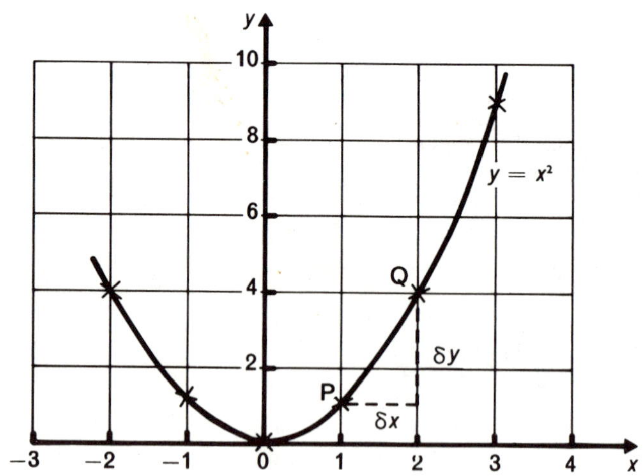

Fig. 1.4

The symbol δx will be used to represent an increment of x, and δy will be used to represent the corresponding increment of y. The gradient of the chord PQ is then

$$\frac{\delta y}{\delta x}$$

When Q is at the point (2, 4) then $\delta x = 1$, and $\delta y = 3$

$$\therefore \qquad \frac{\delta y}{\delta x} = \frac{3}{1} = 3$$

The following table shows how $\frac{\delta y}{\delta x}$ alters as Q moves nearer and nearer to P.

Coordinates of Q				Gradient of
x	y	δy	δx	PQ $= \frac{\delta y}{\delta x}$
2	4	3	1	3
1.5	2.25	1.25	0.5	2.5
1.4	1.96	0.96	0.4	2.4
1.3	1.69	0.69	0.3	2.3
1.2	1.44	0.44	0.2	2.2
1.1	1.21	0.21	0.1	2.1
1.01	1.020 1	0.020 1	0.01	2.01
1.001	1.002 001	0.002 001	0.001	2.001

It will be seen that as Q approaches nearer and nearer to P, the value of $\frac{\delta y}{\delta x}$ approaches 2. It is reasonable to suppose that eventually when Q coincides with P (that is, when the chord PQ becomes a tangent to the curve at P) the gradient of the tangent will be exactly equal to 2. The gradient of the tangent will give us the gradient of the curve at P.

Now as Q approaches P, δx tends to zero and the gradient of the chord, $\frac{\delta y}{\delta x}$, tends, in the limit (as we say), to the gradient of the tangent. We denote the gradient of the tangent as $\frac{dy}{dx}$. We can write all this as

$$\operatorname*{Limit}_{\delta x \to 0} \frac{\delta y}{\delta x} = \frac{dy}{dx}$$

Instead of selecting special values for δy and δx let us now consider the general case, so that P has the coordinates (x, y) and Q has the coordinates

$(x+\delta x, y+\delta y)$, (Fig. 1.5). Q is taken very close to P, so that δx is a very small quantity.

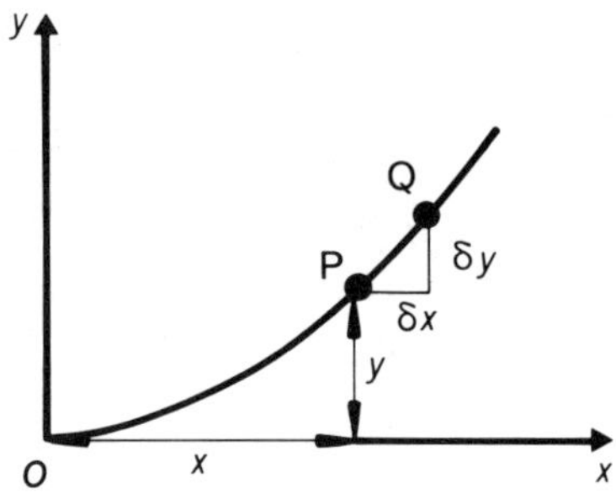

Fig. 1.5

Now:

$$y = x^2$$

and as $Q(x+\delta x, y+\delta y)$ lies on the curve, then

$$y+\delta y = (x+\delta x)^2$$

$$\therefore \quad y+\delta y = x^2+2x\delta x+(\delta x)^2$$

But $y = x^2$, and so we get

$$\delta y = 2x\delta x+(\delta x)^2$$

and, by dividing both sides by δx, the gradient of chord PQ is

$$\frac{\delta y}{\delta x} = 2x+\delta x$$

As Q approaches P, δx tends to zero and $\frac{\delta y}{\delta x}$ tends, in the limit, to the gradient of the tangent of the curve at P. Thus

$$\underset{\delta x \to 0}{\text{Limit}} \frac{\delta y}{\delta x} = \frac{\mathrm{d}y}{\mathrm{d}x} = 2x$$

The process of finding $\frac{\mathrm{d}y}{\mathrm{d}x}$ is called *differentiation.*

The symbol $\frac{\mathrm{d}y}{\mathrm{d}x}$ means the differential coefficient of y with respect to x.

We can now check our assumption regarding the gradient of the curve at P. Since at P the value of $x = 1$, then

$$\frac{dy}{dx} = 2 \times 1 = 2$$

and we see that our assumption was correct.

DIFFERENTIAL COEFFICIENT OF x^n

It can be shown, by a method similar to that used for finding the differential coefficient of x^2, that:

If $$y = x^n$$

then $$\frac{dy}{dx} = nx^{n-1}$$

This is true for all values of n including negative and fractional indices. When we use it as formula it enables us to avoid having to differentiate each time from first principles.

EXAMPLES 4

If $y = x^3$,

then $\dfrac{dy}{dx} = 3x^2$

If $y = \dfrac{1}{x} = x^{-1}$,

then $\dfrac{dy}{dx} = -x^{-2} = -\dfrac{1}{x^2}$

If $y = \sqrt{x} = x^{1/2}$,

then $\dfrac{dy}{dx} = \dfrac{1}{2}x^{-1/2} = \dfrac{1}{2}\dfrac{1}{x^{1/2}} = \dfrac{1}{2\sqrt{x}}$

If $y = \sqrt[5]{x^2} = x^{2/5}$,

then $\dfrac{dy}{dx} = \dfrac{2}{5}x^{2/5-1} = \dfrac{2}{5}x^{-3/5} = \dfrac{2}{5\sqrt[5]{x^3}}$

When a power of x is multiplied by a constant, that constant remains unchanged by the process of differentiation.

Hence if $$y = ax^n$$

then $$\frac{dy}{dx} = anx^{n-1}$$

EXAMPLES 5

If $\quad y = 3x^4,$

then $\quad \dfrac{dy}{dx} = 3\times 4x^3 = 12x^3$

If $\quad y = 2x^{1.3},$

then $\quad \dfrac{dy}{dx} = 2(1.3)x^{0.3} = 2.6x^{0.3}$

If $\quad y = \frac{1}{5}x^7,$

then $\quad \dfrac{dy}{dx} = \frac{1}{5}\times 7x^6 = \frac{7}{5}x^6$

If $\quad y = \frac{3}{4}\sqrt[3]{x} = \frac{3}{4}x^{1/3},$

then $\quad \dfrac{dy}{dx} = \frac{3}{4}\times\frac{1}{3}x^{-2/3} = \frac{1}{4}x^{-2/3}$

If $\quad y = \dfrac{4}{x^2} = 4x^{-2},$

then $\quad \dfrac{dy}{dx} = 4(-2)x^{-3} = -8x^{-3}$

When a numerical constant is differentiated the result is zero. This can be seen since $x^0 = 1$ and we can write, for example, constant 4 as $4x^0$, then differentiating with respect to x we get

$$4(0)x^{-1} = 0$$

If, as an alternative method, we plot the graph of $y = 4$ we get a straight line parallel with the x-axis as shown in Fig. 1.6.

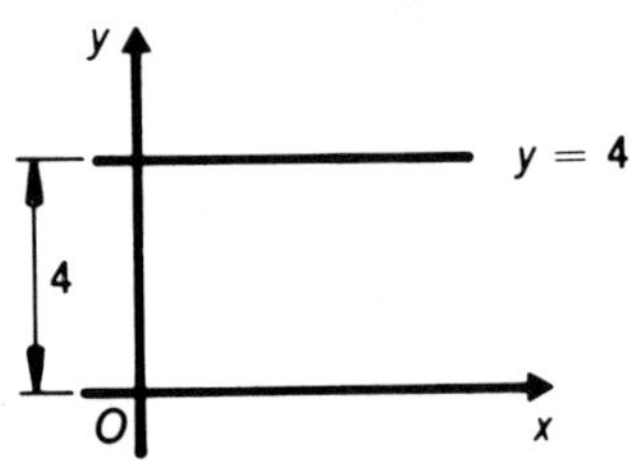

Fig. 1.6

The gradient of the line is zero; that is

$$\frac{dy}{dx} = 0.$$

To differentiate an expression containing the sum of several terms, differentiate each individual term separately.

EXAMPLES 6

If $\quad y = 3x^2+2x+3,$

then $\quad \dfrac{dy}{dx} = 3(2)x+2(1)x^0+0 = 6x+2$

If $y = ax^3+bx^2+cx+d$ where a, b, c and d are constants,

then $\dfrac{dy}{dx} = 3ax^2+2bx+c$

If $y = \sqrt{x}+\dfrac{1}{\sqrt{x}} = x^{1/2}+x^{-1/2}$,

then $\dfrac{dy}{dx} = \tfrac{1}{2}x^{-1/2}+(-\tfrac{1}{2})x^{-3/2} = \dfrac{1}{2\sqrt{x}}-\dfrac{1}{2\sqrt{x^3}}$

If $y = 3.1x^{1.4}-\dfrac{3}{x}+5 = 3.1x^{1.4}-3x^{-1}+5$

then $\dfrac{dy}{dx} = (3.1)(1.4)x^{0.4}-3(-1)x^{-2} = 4.34x^{0.4}+\dfrac{3}{x^2}$

So far our differentiation has been in terms of x and y only. But they are only letters representing variables and we may choose other letters or symbols.

Thus, if $s = 3t^2-1.5t+3$,

then $\dfrac{ds}{dt} = 6t-1.5$

The example which follows shows the application of differentiation for finding the gradient of a curve.

EXAMPLE 7

Find the gradient of the graph $y = 3x^2-3x+4$:
(a) when $x = 3$, and,
(b) when $x = -2$.

The gradient at a point is expressed by $\dfrac{dy}{dx}$.

If $y = 3x^2-3x+4$,

then $\dfrac{dy}{dx} = 6x-3$

When $x = 3$,

$$\frac{dy}{dx} = 6(3)-3 = 15$$

When $x = -2$,

$$\frac{dy}{dx} = 6(-2)-3 = -15$$

Exercise 2

Differentiate the following:

1) $y = x^2$

2) $y = x^7$

3) $y = 4x^3$

4) $y = 6x^5$

5) $s = 0.5t^3$

6) $A = \pi R^2$

7) $y = x^{1/2}$

8) $y = 4x^{3/2}$

9) $y = 2\sqrt{x}$

10) $y = 3\sqrt[3]{x^2}$

11) $y = \dfrac{1}{x^2}$

12) $y = \dfrac{1}{x}$

13) $y = \dfrac{3}{5x}$

14) $y = \dfrac{2}{x^3}$

15) $y = \dfrac{1}{\sqrt{x}}$

16) $y = \dfrac{2}{3\sqrt{x}}$

17) $y = \dfrac{5}{x\sqrt{x}}$

18) $s = \dfrac{3\sqrt{t}}{5}$

19) $K = \dfrac{0.01}{h}$

20) $y = \dfrac{5}{x}$

21) $y = 4x^2 - 3x + 2$

22) $s = 3t^3 - 2t^2 + 5t - 3$

23) $q = 2u^2 - u + 7$

24) $y = 5x^4 - 7x^3 + 3x^2 - 2x + 5$

25) $s = 7t^5 - 3t^2 + 7$

26) $y = \dfrac{x + x^3}{\sqrt{x}}$

27) $y = \dfrac{3 + x^2}{x}$

28) $y = \sqrt{x} + \dfrac{1}{\sqrt{x}}$

29) $y = x^3 + \dfrac{3}{\sqrt{x}}$

30) $s = t^{1.3} - \dfrac{1}{4t^{2.3}}$

31) $y = \dfrac{3x^3}{5} - \dfrac{2x^2}{7} - \sqrt{x}$

32) $y = 0.08 + \dfrac{0.01}{x}$

33) $y = 3.1x^{1.5} - 2.4x^{0.6}$

34) $y = \dfrac{x^3}{2} - \dfrac{5}{x} + 3$

35) $s = 10 - 6t + 7t^2 - 2t^3$

36) Find the gradient of the curve $y = 3x^2 + 7x + 3$ at the points where $x = -2$ and $x = 2$.

37) Find the gradient of the curve $y = 2x^3 - 7x^2 + 5x - 3$ at the points where $x = -1.5$, $x = 0$ and $x = 3$.

38) Find the gradient of the graph $2x^2 - 5$ at the points where $x = 1$ and $x = +2$.

39) Find the gradient of the curve $y = x^2 - 3x + 2$ at the point where $x = 3$.

40) For what values of x is the gradient of the curve

$$y = \frac{x^3}{3} + \frac{x^2}{2} - 33x + 7$$

equal to 3?

41) For what values of x is the gradient of the curve $y = 3 + 4x - x^2$ equal to (a) -1, (b) 0, (c) 2?

42) If $y = x - \dfrac{1}{x}$ find $\dfrac{dy}{dx}$ where $x = 1$.

TO FIND THE RATE OF CHANGE OF sin θ

The rate of change of a curve at any point is the gradient of the tangent at that point. We shall, therefore, find the gradient at various points on the graph of $\sin\theta$ and then plot the values of these gradients to obtain a new graph.

It is suggested that the reader follows the method given, plotting his own curves on graph paper.

First, we plot the graph of $y = \sin\theta$ from $\theta = 0°$ to $\theta = 90°$ using values of $\sin\theta$ obtained from tables which are:

θ	0°	15°	30°	45°	60°	75°	90°
$y = \sin\theta$	0	0.259	0.500	0.707	0.866	0.966	1.000

These values are shown plotted in Fig. 1.7.

Consider point P on the curve, where $\theta = 45°$, and draw the tangent APM. As shown in Example 1 of this chapter, we can find the gradient of the tangent by constructing a suitable right-angled triangle AMN (which should be as large as conveniently possible for accuracy) and finding the value of $\frac{MN}{AN}$.

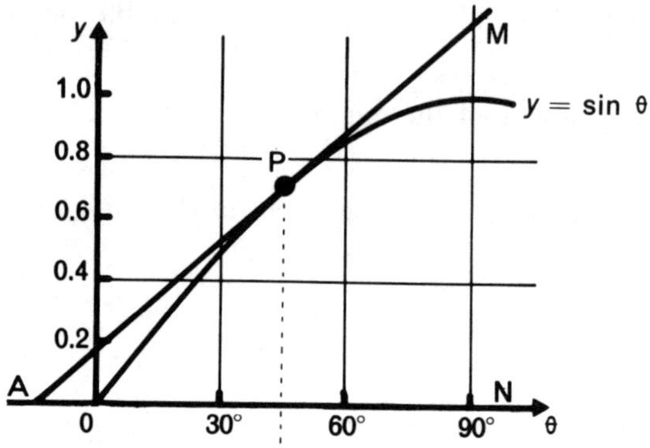

Fig. 1.7

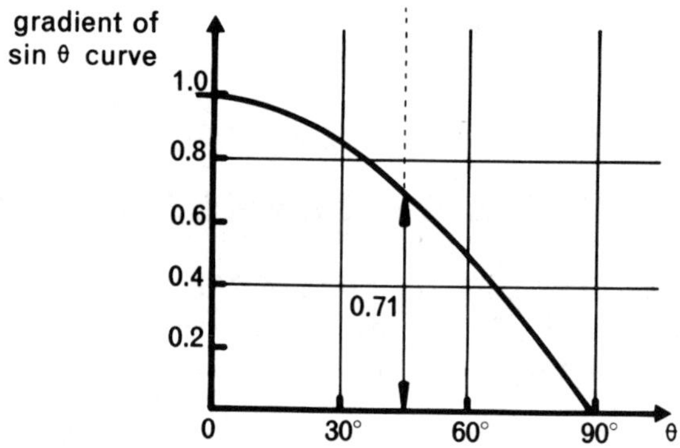

Fig. 1.8

Using the scale on the y-axis MN = 1.29 by measurement and using the scale on the θ-axis AN = 104° by measurement.

In calculations of this type it is necessary to measure AN in radians.

Remember that $\quad 360° = 2\pi$ radians

$$\therefore \quad 1° = \frac{2\pi}{360} \text{ radians}$$

$$\therefore \quad 104° = \frac{2\pi}{360} \times 104 = 1.81 \text{ radians}$$

Hence, $\quad$ the gradient at P $= \frac{MN}{AN} = \frac{1.29}{1.81} = 0.71$

The value 0.71 is used as the y-value at $\theta = 45°$ to plot a point on a new graph using the same scales as before. This new graph could be plotted on the same axes as $y = \sin\theta$ but for clarity it has been shown on new axes shown in Fig. 1.7.

This procedure is repeated for points on the $\sin\theta$ curve at θ values 0°, 15°, 30°, 60°, 75° and 90° and the new curve obtained will be as shown in Fig. 1.8. This is the graph of the gradients of the sine curve at various points.

If we now plot a graph of $\cos\theta$, taking values from tables, on the axes in Fig. 1.8 we shall find that the two curves coincide — any difference will be due to errors from drawing the tangents.

Hence the gradient of the $\sin\theta$ curve at any value of θ is the same as the value of $\cos\theta$.

In other words the rate of change of $\sin\theta$ is $\cos\theta$, and in mathematical notation:

If $$y = \sin\theta$$

then $$\frac{dy}{dx} = \cos\theta$$

In the above work we have only considered the graphs between 0° and 90° but the results are true for all values of the angle.

Similarly it may be shown that:

If $$y = \cos\theta$$

then $$\frac{dy}{dx} = -\sin\theta$$

MAXIMUM AND MINIMUM VALUES OF A CURVE

Consider the equation:

$$y = x^2 - 8x + 15$$

To draw a graph of this equation we draw up a table in the usual way:

x	0	1	2	3	4	5	6	7	8
x^2	0	1	4	9	16	25	36	49	64
$-8x$	0	-8	-16	-24	-32	-40	-48	-56	-64
$+15$	15	15	15	15	15	15	15	15	15
$y = x^2-8x+15$	15	8	3	0	-1	0	3	8	15

From Fig. 1.9 we see that at $x = 4$ the curves reaches its lowest value. This point is called a *turning point*.

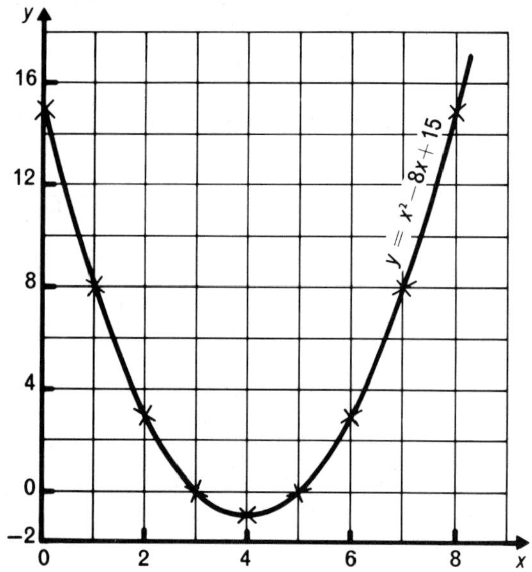

Fig. 1.9

Now consider the equation

$$y = -x^2+8x+15$$

Drawing up a table we get

x	0	1	2	3	4	5	6	7	8
$y = -x^2+8x+15$	15	22	27	30	31	30	27	22	15

From Fig. 1.10 we see that at the point $x = 4$ the graph reaches its highest value, this point is also a *turning point*.

The curves $y = x^2-8x+15$ and $y = -x^2+8x+15$ have different types of turning point.

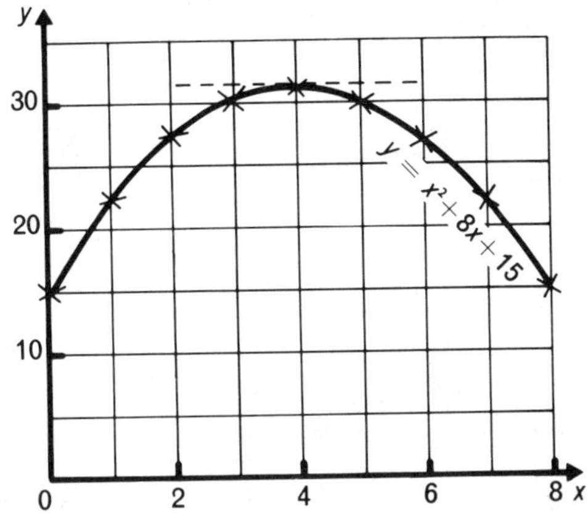

Fig. 1.10

(i) In $y = x^2 - 8x + 15$ the turning point is the lowest point on the curve. The values of y close to the turning point and on either side of it are greater than at the turning point. The value of y at such a turning point is called a *minimum value.*

(ii) In $y = -x^2 + 8x + 15$ the turning point is the highest point on the curve. We see that the values of y just before the turning point and just after it are less than the value of y at the turning point itself. The value of y at a turning point like this is called a *maximum value.*

We can see, also, that at these maximum and minimum points the gradient of the curve is zero, that is, the rate of change is zero. This fact is of great importance and will be used many times in your future studies.

EXAMPLE 8

Plot the graph of $y = x^3 - 5x^2 + 2x + 8$ for values of x between -2 and 6. Hence find the maximum and minimum values of y.

We draw up a table as before:

x	-2	-1	0	1	2	3	4	5	6
$y = x^3 - 5x^2 + 2x + 8$	-24	0	8	6	0	-4	0	18	56

The graph is shown in Fig. 1.11. The maximum value occurs at the point P and the minimum value at the point Q. The maximum value of y is 8.21 and the minimum value of y is -4.06.

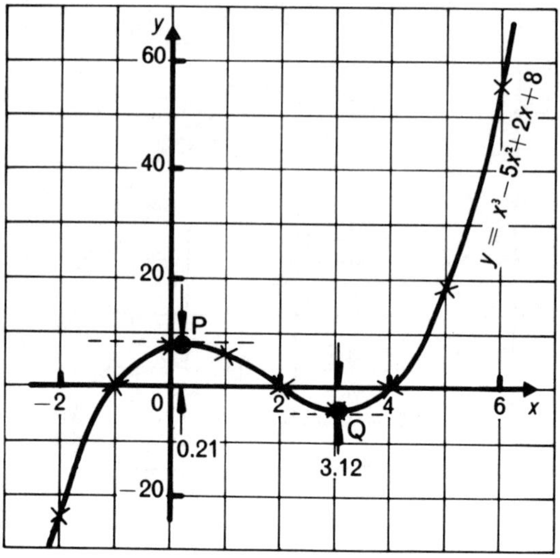

Fig. 1.11

Notice that the value of y at the point P is not the greatest value of y nor is the value of y at point Q the least. However the values of y at P and Q are called maximum and minimum values of y.

Note also that at both the points P and Q the gradient of the curve is zero.

Exercise 3

Differentiate the expressions in Questions 1–10 each with respect to the appropriate variable.

1) $y = 4 \sin x$

2) $y = 4 + \sin x$

3) $y = x - 4 \sin x$

4) $y = 3 \sin x - 2$

5) $i = 3 \cos \theta$

6) $v = t + 4 \cos t$

7) $y = \theta^2 - 2 \cos \theta$

8) $i = 2 \cos \theta + 3 \sin \theta$

9) $h = 4t^{2.4} - \dfrac{\cos t}{4} + 16$

10) $p = \dfrac{10}{v} + 6 \sin v$

11) Show that the following expressions never decrease:

(a) $y = 2x + \cos x$ (b) $i = \sin t + t$

12) Find the gradient of the curve $y = \cos\theta + \sin\theta$ at each of the following points.

(a) $\theta = 0$ (b) $\theta = \frac{\pi}{4}$ (c) $\theta = \pi$

13) Find the minimum value of the curve $y = 3x^2 - 2x + 3$. Plot the graphs for values of x between -2 and 3.

14) Find the maximum value of the curve $y = -x^2 + 5x + 7$. Plot the graph for values of x between -2 and 4.

15) Find the maximum and minimum values of $y = 2x^3 - 24x$. Plot the graph for values of x between -4 and 4.

16) In an electric circuit the number of heat units, H, produced when a current, i, is flowing is given by $H = Ei - Ri^2$ where E is the e.m.f. and R is the resistance. Find the maximum heat that can be produced when $E = 8$ and $R = 2$.

FUNCTIONS

When two variable quantities, x and y, are so related that each value of x is related to exactly one value of y, thus the value of y depends on the value of x, then y is said to be a **function** of x.

We can write this as $y = f(x)$.

When $y = 3x^2 + 4x - 7$ and $x = 5$ then $y = 3(5^2) + 4(5) - 7 = 88$.

Since the value of y depends on the value allocated to x, y is a function of x.

Also $\frac{dy}{dx} = 6x + 4$. The expression $6x + 4$ is another function of x known as the **first derived function** or the **first derivative.** We can write

$$f(x) = 3x^2 + 4x - 7$$

$$\Rightarrow \qquad f'(x) = 6x + 4$$

If
$$y = f(x)$$
then
$$\frac{dy}{dx} = f'(x)$$

Both methods of indicating a derivative are used.

Thus if $i = 4t^2 + 10$ we say that i is a function of t. That is the value of i depends on the value chosen for t.

Hence $$i = f(t)$$

and $$\frac{di}{dt} = 8t \quad \text{or} \quad f'(i) = 8t$$

FUNCTION OF A FUNCTION

A function like $y = (2x-6)^3$ is a **function of a function** because $(2x-6)$ is a function of x and $(2x-6)^3$ is a function of $(2x-6)$. Another example of a function of a function is $i = 20 \sin(20t-0.3)$ where i is a function of the sine of an angle which itself is a function of time.

To differentiate a function of a function it is often helpful if we substitute a single variable for the 'inner' function.

DIFFERENTIATION OF A FUNCTION OF A FUNCTION BY SUBSTITUTION

The letter U is often substituted for the 'inner' function when differentiating a function of a function together with the formula:

$$\boxed{\frac{dy}{dx} = \frac{dy}{dU} \times \frac{dU}{dx}}$$

The proof of this important formula is beyond the scope of this course but the application is straightforward and is best shown by example.

EXAMPLE 9

Find $\dfrac{dy}{dx}$ if $y = (x^2-x)^9$.

We have $$y = (x^2-x)^9$$

then $$y = U^9 \qquad \text{where } U = x^2-x$$

$$\Rightarrow \quad \frac{dy}{dU} = 9U^8 \qquad \text{and} \quad \frac{dU}{dx} = 2x-1$$

but $$\frac{dy}{dx} = \frac{dy}{dU} \times \frac{dU}{dx}$$

$$\Rightarrow \quad \frac{dy}{dx} = 9U^8 \times (2x-1)$$

The differentiation has now been completed and it only remains to replace U in terms of x by using our original substitution $U = x^2 - x$.

Hence $$\frac{dy}{dx} = 9(x^2-x)^8(2x-1)$$

EXAMPLE 10

Find $\frac{di}{dt}$ if $i = 200 \sin(20t-0.3)$.

We have $$i = 200 \sin(20t-0.3)$$

hence $$i = 200 \sin U \qquad \text{where } U = 20t-0.3$$

$$\Rightarrow \qquad \frac{di}{dU} = 200 \cos U \qquad \text{and } \frac{dU}{dt} = 20$$

but $$\frac{di}{dt} = \frac{di}{dU} \times \frac{dU}{dt}$$

$$\Rightarrow \qquad \frac{di}{dt} = 200 \cos U \times 20$$

Replacing U gives

$$\frac{di}{dt} = 4000 \cos(20t-0.3)$$

DIFFERENTIATION OF A FUNCTION OF A FUNCTION BY RECOGNITION

Consider $y = (\ )^n$ where any function of x can be written inside the bracket. Differentiating with respect to x gives

$$\frac{dy}{dx} = \frac{dy}{d(\ )} \times \frac{d(\ )}{dx}$$

Thus to differentiate an expression of the type $(\ )^n$ we first differentiate the bracket, treating it as a term similar to x^n. Secondly differentiate the function of x inside the bracket. Finally, to obtain an expression for $\frac{dy}{dx}$, multiply the two results together.

EXAMPLE 11

Determine $\frac{dy}{dx}$ if $y = (x^2-5x+3)^5$.

Differentiating the bracket as a whole we have

$$\frac{dy}{d(\ )} = 5(x^2-5x+3)^4 \qquad [1]$$

Differentiating the function inside the bracket, x^2-5x+3, gives

$$\frac{d(\ )}{dx} = 2x-5 \qquad [2]$$

Multiplying the results [1] and [2] together gives

$$\frac{dy}{dx} = 5(x^2-5x+3)^4\times(2x-5)$$

$$= 5(2x-5)(x^2-5x+3)^4$$

Hence by recognising the method we can differentiate directly.

Consider for example $\quad y = (x^2-3x)^7$

Hence $$\frac{dy}{dx} = 7(x^2-3x)^6\times(2x-3)$$

$$= 7(2x-3)(x^2-3x)^6$$

EXAMPLE 12

Find $\dfrac{d\theta}{dt}$ if $\theta = 5\cos\left(2t^2+\dfrac{3\pi}{2}\right)$.

In this example the function of a function is composed of a trigonometric function, 5 cos ( ), and an algebraic function, $\left(2t^2+\dfrac{3\pi}{2}\right)$.

We differentiate the two functions in this order and then multiply the two results together to get the required expression for $\dfrac{d\theta}{dt}$.

Thus $$\frac{d\theta}{dt} = -5\sin\left(2t^2+\frac{3\pi}{2}\right)\times 4t$$

$$= -20t\sin\left(2t^2+\frac{3\pi}{2}\right)$$

EXAMPLE 13

Determine $f'(\theta)$ if $f(\theta) = \sin 7\theta$.

{$f'(\theta)$ = first derived trigonometric function × first derived algebraic function.}

Hence $$f'(\theta) = \cos 7\theta\times 7$$

$$= 7\cos 7\theta$$

Exercise 4

In each of the following questions find the differential coefficient with respect to the appropriate variable.

1) $y = (x+1)^5$

2) $y = (3x+2)^3$

3) $y = (6+x)^4$

4) $y = (4x-1)^7$

5) $y = (3-x)^3$

6) $y = (2-3x)^3$

7) $y = (2-x)^4$

8) $y = (4-2x)^5$

9) $y = (x^2-1)^4$

10) $y = (1+x^3)^3$

11) $y = (x^4-2)^5$

12) $y = (1-x^5)^2$

13) $i = (2t+3)^{1/2}$

14) $v = (3\theta-1)^{1/3}$

15) $A = (r-1)^{-4}$

16) $\theta = (2-3t)^{-2}$

17) $r = \sqrt{(2h-5)}$

18) $h = \sqrt{(1-4r)}$

19) $y = \dfrac{6}{1-x}$

20) $y = \dfrac{4}{(2x+7)^2}$

21) $y = \sin 4x$

22) $y = \sin(4x+3)$

23) $i = 5\sin(2\pi t-0.04)$

24) $y = \cos 2x$

25) $y = \cos(2x+3)$

26) $v = 40\cos(100\pi t)$

27) $i = 5\cos(100\pi t-0.001)$

28) $i = 5\cos(3t-0.1)$

29) $v = 250\sin(2\pi t)$

30) $v = 50\cos(0.05-2\pi t)$

PROJECTS

1) Investigate relationships like $\dfrac{dM}{dx} = F$, which exist between Intensity of Loading, Shearing Force and Bending Moment of a beam carrying a uniformly distributed load of w per unit length.

Cantilevers should also be considered.

2) Investigate the formation of Difference Tables. Using appropriate values from, say Steam Tables, form a difference table and show the relationship between first differences and the differential coefficient.

2. THE BINOMIAL THEOREM

On reaching the end of this chapter you should be able to:-

1. *Expand expressions of the form* $(a+x)^n$ *for small, positive integer* n.
2. *State the general form for the binomial coefficients for all* n.
3. *Solve problems using the binomial theorem relevant to problems in technology, to include approximations.*

BINOMIAL EXPRESSION

A **binomial expression** consists of two terms. Thus $1+x$, $a+b$, $5y-2$, $3x^2+7$ and $7a^3+3b^2$ are all binomial expressions. The **binomial theorem** allows us to expand powers of such expressions.

THE BINOMIAL THEOREM

Now $\qquad (a+b)^0 = 1$

(since any number to the power 0 is unity),

and also $\qquad (a+b)^1 = a+b$

By multiplying, $\qquad (a+b)^2 = a^2+2ab+b^2$

also $\qquad (a+b)^3 = a^3+3a^2b+3ab^2+b^3$

and $\qquad (a+b)^4 = a^4+4a^3b+6a^2b^2+4ab^3+b^4$

We can now arrange the coefficients of each of the terms of the above expansions in the form known as **Pascal's triangle**

Binomial expression	Coefficients in the expansion
$(a+b)^0$	1
$(a+b)^1$	1 1
$(a+b)^2$	1 2 1
$(a+b)^3$	1 3 3 1
$(a+b)^4$	1 4 6 4 1
$(a+b)^5$	1 5 10 10 5 1
$(a+b)^6$	1 6 15 20 15 6 1
$(a+b)^7$	1 7 21 35 35 21 7 1

It will be seen that:

(a) The number of terms in each expansion is one more than the index. Thus the expansion of $(a+b)^9$ will have 10 terms.

(b) The arrangement of the coefficients is symmetrical.

(c) The coefficients of the first and last terms are both always unity.

(d) Each coefficient in the table is obtained by adding together the two coefficients in the line above which lie on either side of it.

The expansion of $(a+b)^8$ is therefore:

$$a^8+(1+7)a^7b+(7+21)a^6b^2+(21+35)a^5b^3+(35+35)a^4b^4$$
$$+(35+21)a^3b^5+(21+7)a^2b^6+(7+1)ab^7+b^8$$
$$=a^8+8a^7b+28a^6b^2+56a^5b^3+70a^4b^4+56a^3b^5+28a^2b^6+8ab^7+b^8$$

It is inconvenient to use Pascal's traingle when expanding higher powers of $(a+b)$. In such cases the following series is used.

$$(a+b)^n = a^n+na^{n-1}b+\frac{n(n-1)}{2!}a^{n-2}b^2+\frac{n(n-1)(n-2)}{3!}a^{n-3}b^3+\ldots+b^n$$

This is the **binomial theorem** and is true for all positive whole numbers n.

The symbol ! indicates 'factorial' when following a positive whole number.

For example, 4! is pronounced 'factorial four' and means $4\times3\times2\times1$. Thus $2! = 2\times1$, $3! = 3\times2\times1$, $4! = 4\times3\times2\times1$, $5! = 5\times4\times3\times2\times1$, and so on.

EXAMPLE 1

Expand $(3x+2y)^4$.

Comparing $(3x+2y)^4$ with $(a+b)^4$, we have $3x$ in place of a, and $2y$ in place of b. Substituting in the standard expansion, we get:

$$[(3x)+(2y)]^4 = (3x)^4+4(3x)^3(2y)+\frac{(4)(4-1)}{2!}(3x)^2(2y)^2 + \frac{(4)(4-1)(4-2)}{3!}(3x)(2y)^3+(2y)^4$$

$$= (3x)^4+4(3x)^3 2y+\frac{4\times 3}{2\times 1}(3x)^2(2y)^2+\frac{4\times 3\times 2}{3\times 2\times 1}(3x)(2y)^3+(2y)^4$$

$$= 81x^4+216x^3y+216x^2y^2+96xy^3+16y^4$$

EXAMPLE 2

Find the first four terms of the expansion of $(x-4y)^{15}$.

The given binomial expression should be rewritten as $[x+(-4y)]^{15}$ and comparing this with the standard expression for $(a+b)^n$ we have x in place of a, $-4y$ in place of b, and 15 in place of n.

$$\therefore\ [x+(-4y)]^{15} = x^{15}+15x^{14}(-4y)+\frac{(15)(15-1)}{2!}x^{13}(-4y)^2 + \frac{15(15-1)(15-2)}{3!}x^{12}(-4y)^3$$

$$= x^{15}+15(-4)x^{14}y+\frac{15\times 14\times(-4)^2}{2\times 1}x^{13}y^2 + \frac{15\times 14\times 13\times(-4)^3}{3\times 2\times 1}x^{12}y^3$$

$$= x^{15}-60x^{14}y+1680x^{13}y^2-29\,120x^{12}y^3$$

APPLICATION TO SMALL ERRORS

In the binomial theorem if we put $a = 1$ and $b = x$ we get

$$(1+x)^n = 1+nx+\frac{n(n-1)}{2!}x^2+\frac{n(n-1)(n-2)}{3!}x^3+\dots$$

When x is very small,

$$\mathbf{(1+x)^n = 1+nx} \text{ approximately,}$$

as all other terms in the series contain powers of x which are negligible when compared with the first two shown. This relationship is often useful.

It is easy enough to measure the expansion of a metal rod which has been heated. It is, however, not so easy to measure the increases in areas and volumes which have been expanded.

If α is the coefficient of linear expansion, then the length of an expanded bar originally 1 m long is $(1+\alpha)$ m for one degree temperature increase. An area of the same material originally $1\text{ m}\times 1\text{ m}$ becomes $(1+\alpha)^2\text{ m}^2$ when it expands. From the above approximation as α is very small then:

$$(1+\alpha)^2 = 1+2\alpha$$

Hence *the area coefficient is approximately twice the linear coefficient.*

Similarly a volume $1\text{ m}\times 1\text{ m}\times 1\text{ m}$ on expanding becomes $(1+\alpha)^3\text{ m}^3$ and again, using the above approximation:

$$(1+\alpha)^3 = 1+3\alpha$$

which shows that *the volume coefficient is approximately three times the linear coefficient.*

EXAMPLE 3

In measuring the radius of a circle, the measurement is 1% too large. If this measurement is used to calculate the area of the circle find the resulting error.

Let A be the area of the circle, and r the radius. If δA is the error in the area, then

$$A = \pi r^2$$

and

$$A+\delta A = \pi(r+r\times\tfrac{1}{100})^2$$
$$= \pi r^2(1+\tfrac{1}{100})^2$$

and since $\frac{1}{100}$ is small when compared with 1, then

$$A+\delta A = \pi r^2(1+2\times\tfrac{1}{100})$$
$$= A(1+\tfrac{2}{100})$$
$$\delta A = \tfrac{2}{100}A$$

Hence

$$\delta A = 2\% \text{ of } A$$

Thus an error of 1% too large in the radius gives an error of 2% too large in the area.

EXAMPLE 4

When a uniform beam is simply supported at its ends the deflection at the centre of span is given by

$$y = \frac{5wl^4}{384EI}$$

where w is the distributed load per unit length, l is the length between supports, E is Young's modulus and I is the 2nd moment of cross-sectional area. Find the percentage change in y when l decreases by 3%.

Let the change in y be δy. Then:

$$y+\delta y = \frac{5w}{384EI}[l-l\times\tfrac{3}{100}]^4$$
$$= \frac{5wl^4}{384EI}[1+(-\tfrac{3}{100})]^4$$
$$= \frac{5wl^4}{384EI}[1+4(-\tfrac{3}{100})] \quad \text{approximately}$$

since all higher powers of $\frac{3}{100}$ are negligible:

Thus $\qquad y+\delta y = y[1-\frac{12}{100}]$

$\therefore \qquad \delta y = -\frac{12}{100}y \quad$ or $\quad 12\%$ of y (decrease)

Hence y will decrease by 12% approximately.

EXAMPLE 5

A formula used in connection with close-coiled helical springs is:

$$P = \frac{GFd^5}{8hD^3}$$

Find the change in P if d is increased by 2% and D is decreased by 3%. Let δP be the change in P. Then:

$$P+\delta P = \frac{GF[d+\frac{2}{100}d]^5}{8h[D-\frac{3}{100}D]^3}$$
$$= \frac{GFd^5[1+\frac{2}{100}]^5}{8hD^3[1-\frac{3}{100}]^3}$$
$$= P[1+(\tfrac{2}{100})]^5[1+(-\tfrac{3}{100})]^{-3}$$
$$= P[1+5(\tfrac{2}{100})][1+(-3)(-\tfrac{3}{100})]$$

since all higher powers of $\frac{2}{100}$ and $\frac{3}{100}$ are negligible.

Thus $\qquad P+\delta P = P[1+\frac{10}{100}][1+\frac{9}{100}]$

$$= P[1+\tfrac{19}{100}] \quad \text{approximately}$$
$$= P+\tfrac{19}{100}P$$

$\therefore \qquad \delta P = \frac{19}{100}P = 19\%$ of P

Hence P will increase by 19% approximately.

Exercise 5

Expand using Pascal's triangle:

1) $(1+z)^5$ **3)** $(x-3y)^4$ **5)** $(2x+y)^7$

2) $(p+q)^6$ **4)** $(2p-q)^5$ **6)** $\left(x+\frac{1}{x}\right)^3$

Find the first four terms of the following expansions using the binomial theorem formula:

7) $(1+x)^{12}$ **9)** $(p+q)^{16}$ **11)** $(x^2-3y)^9$

8) $(1-2x)^{14}$ **10)** $(1+3y)^{10}$ **12)** $\left(x^2+\frac{1}{x^2}\right)^{11}$

13) Show that an error of 1% in the measurement of the radius of a sphere leads to an error of approximately 2% in the outer surface area, and 3% in the volume.

14) Find the approximate percentage error in the calculated volume of a right circular cone if the radius is taken as 2% too large, and the height is taken as 3% too small.

15) A formula used in connection with helical springs is $y = \frac{8WnD^3}{Gd^4}$.

Find the percentage error in y if D is 1% too small, and d is 2% too large.

16) In the standard gas equation $\frac{pv}{T} = k$, the volume v is increased by 2%, the temperature T is diminished by 1%, and the value of constant k remains unaltered. What is the corresponding percentage change in the pressure p?

17) The deflection y at the centre of a steel rod of length l and circular cross-section of diameter d, simply supported at its ends and carrying a concentrated load W at its centre is given by the formula $y = \frac{Wl^3}{d^4}$

Find the percentage change in y when l increases by 2% and d decreases by 3%.

THE BINOMIAL SERIES FOR ANY INDEX NUMBER

When the index n of $(1+x)^n$ is negative or fractional we still obtain the series,

$$(1+x)^n = 1+nx+\frac{n(n-1)}{2!}x^2+\frac{n(n-1)(n-2)}{3!}x^3+\ \ldots$$

with two important differences in the result.

The series is never-ending — that is, it tends to infinity — and secondly it will only converge on a sensible answer if when x is replaced by a number then $-1<x<1$.

Hence:

(a) $(a+b)^n = a^n\left(1+\dfrac{b}{a}\right)^n$

(b) $(1+x)^n = 1+nx+\dfrac{n(n-1)}{2!}x^2+\dfrac{n(n-1)(n-2)}{3!}x^3+\ \ldots$

(i) If $n = 1, 2, 3, \ldots$ then the series will have 2, 3, 4, . . . terms and x can be replaced by any number.

(ii) If n is negative or fractional, or both, the series will have an infinite number of terms which will converge on the answer only if $-1<x<1$.

EXAMPLE 6

Expand $\dfrac{1}{(1+x)^2}$ as far as the term in x^3 giving the limits for which the expansion is valid.

$$\frac{1}{(1+x)^2} = (1+x)^{-2}$$

$$= 1+(-2)x+\frac{-2(-2-1)}{2!}x^2+\frac{-2(-2-1)(-2-2)}{3!}x^3+\ \ldots$$

$$= 1-2x+3x^2-4x^3+\ \ldots$$

This expansion is valid for any value of x between $x = -1$ and $x = 1$. That is for $-1<x<1$.

EXAMPLE 7

Replace $\sqrt{(4-x)}$ with an equivalent expression and hence expand it as a series as far as the term in x^3. State the range of values of x which when substituted into the series will give a convergent answer. Use the series to determine an approximate value to $\sqrt{3}$.

$$\sqrt{(4-x)} = (4-x)^{1/2} = 4^{1/2}\left(1-\frac{x}{4}\right)^{1/2} = 2\left(1-\frac{x}{4}\right)^{1/2}$$

On expanding $\left(1-\dfrac{x}{4}\right)^{1/2}$ the resulting series will be valid for $-1<\dfrac{x}{4}<1$.

That is for $-4<x<4$.

Therefore,

$$2\left(1-\frac{x}{4}\right)^{1/2} = 2\left[1+\frac{1}{2}\left(-\frac{x}{4}\right)+\frac{\frac{1}{2}(\frac{1}{2}-1)}{2!}\left(-\frac{x}{4}\right)^2+\frac{\frac{1}{2}(\frac{1}{2}-1)(\frac{1}{2}-2)}{3!}\left(-\frac{x}{4}\right)^3+\ldots\right]$$

$$= 2\left[1-\frac{1}{8}x-\frac{1}{128}x^2-\frac{1}{1024}x^3-\ \ldots\right]$$

$$\sqrt{(4-x)} = 2-\frac{1}{4}x-\frac{1}{64}x^2-\frac{1}{512}x^3-\ \ldots \quad \text{where } -4<x<4.$$

Replacing x by 1, which is within the range of valid replacements, we obtain

$$\sqrt{3} = 2-\frac{1}{4}-\frac{1}{64}-\frac{1}{512}-\ \ldots \approx 1.732$$

EXAMPLE 8

Show that if x is a small number then $\dfrac{\sqrt{(1+2x)}}{(3-x)^2} \approx 3+5x.$

$$\frac{\sqrt{(1+2x)}}{(3-x)^2} = (1+2x)^{1/2}(3-x)^{-2}$$

and using the approximation $(1+x)^n = 1+nx$

$$\approx (1+x)(3+2x) \approx 3+5x,$$

ignoring squares and higher powers of x.

Exercise 6

Rewrite the expressions in Questions 1–4 in the form $a^n\left(1+\dfrac{b}{a}\right)^n$.

1) $(3-2x)^{-2}$ **2)** $(16+x)^{1/2}$ **3)** $\dfrac{1}{\sqrt{(9-x)}}$ **4)** $(8+2.4x)^{2/3}$

For the expressions in Questions 5–8 determine the first four terms of their series expansion and in each case state the restrictions on the values of x.

5) $(1-2x)^{-3}$ **6)** $(4+x)^{-1/2}$ **7)** $\dfrac{1}{\sqrt{(1-2x)}}$ **8)** $\sqrt{(9-1.8x)}$

9) Obtain a series expansion for $(1+x)^{1/2}$ as far as the term in x^3. By rewriting 1.05 as $(1+0.05)$ use the series expansion to determine $\sqrt{1.05}$ to six places of decimals. Use a calculator to check your answer.

10) Given that x is so small that terms involving x^2 and higher powers can be ignored show that

(a) $(1-x)^{1/3}(1+2x)^{-3/4} \approx 1-\dfrac{11}{6}x$ (b) $\dfrac{(1-\frac{1}{2}x)^2(1+2x)^{1/2}}{(1+3x^2)^{1/3}} \approx 1$

(c) $\dfrac{1}{(1-2x)^2(1-3x)^3} \approx 1+13x$

3. THE EXPONENTIAL FUNCTION AND LOGARITHMS

After reaching the end of this chapter you should be able to:-

1. *State the laws of indices:*
   $a^m \times a^n = a^{m+n}$, $\dfrac{a^m}{a^n} = a^{m-n}$, $(a^m)^n = a^{mn}$,
   $\sqrt[m]{a^n} = a^{n/m}$ $\dfrac{1}{a^n} = a^{-n}$, $a^0 = 1$.
2. *Solve equations containing logarithms.*
3. *State the rules for the use of logarithms.*
4. *Solve indicial equations.*
5. *State the expansion of the exponential function in a power series.*
6. *Plot the graphs of the functions* $y = e^x$ *and* $y = e^{-x}$*, using tables.*
7. *Define Napierian logarithms.*
8. *Determine Napierian logarithms from tables.*

LAWS OF INDICES

Multiplication: $a^m \times a^n = a^{m+n}$.

Division: $\dfrac{a^m}{a^n} = a^{m-n}$.

Powers: $(a^m)^n = a^{mn}$.

Roots: $\sqrt[m]{a^n} = a^{n/m}$.

Reciprocals: $\dfrac{1}{x^n} = x^{-n}$.

Zero index: $a^0 = 1$.

EXAMPLE 1

a) $a^3 \times a^4 \times a^7 = a^{3+4+7} = a^{14}$.

b) $\dfrac{b^4c^6}{b^2c^9} = \dfrac{b^{4-2}}{c^{9-6}} = \dfrac{b^2}{c^3}$.

c) $\sqrt[3]{27a^6} = (3^3a^6)^{1/3} = 3^{3\times1/3}a^{6\times1/3} = 3a^2$.

d) $(2x^2)^{-3} = \dfrac{1}{(2x^2)^3} = \dfrac{1}{2^3x^6}$

e) $37^0 = 1$

Exercise 7

Simplify each of the following with positive indices:

1) $a^5 \times a^6$ **2)** $z^4 \times z^7$ **3)** $y^3 \times y^4 \times y^5$

4) $2^3 \times 2^5$

5) $3 \times 3^2 \times 3^5$

6) $\frac{1}{2}a \times \frac{1}{4}a^2 \times \frac{3}{4}a^3$

7) $a^5 \div a^2$

8) $m^{12} \div m^5$

9) $2^8 \div 2^4$

10) $x^{20} \div x^5$

11) $a^5 \times a^3 \div a^4$

12) $q^7 \times q^6 \div q^5$

13) $\frac{m^5}{m^3} \times \frac{m}{m^2}$

14) $\frac{l^5 \times l^6}{l^2 \times l^7}$

15) $\frac{aL^4}{aL^2}$

16) $(x^3)^4$

17) $(a^5)^3$

18) $(3x^4)^2$

19) $(2^3)^2$

20) $(10^3)^2$

21) $(ab^2)^3$

22) $(ab^2c^3)^4$

23) $(2x^2y^3z)^5$

24) $\left(\frac{3m^2}{4n^3}\right)^5$

25) Find the values of 10^{-1}, 2^{-2}, 3^{-4}, 5^{-2}.

26) Find the values of $2^4 \times 2$, $5^2 \times 5^{1/2} \times 5^{3/2}$, $8^{1/3}$, $27^{1/3}$.

27) Express as powers of 3: 9^3, 27^5, 81^3, $9^4 \times 27^3$.

28) Express as powers of a: $\sqrt[5]{a}$, $\sqrt[3]{a^2}$, $\sqrt[7]{a^4}$, $\sqrt{a^6}$.

29) Find the values of: $2^4 \times 2^2$, $(10^2)^3$, $(2^5)^{2/5}$, $64^{1/2}$.

30) Evaluate: $32^{2/5}$, $16^{3/4}$, $(81)^{-1/4}$, $9^{-1/2}$.

31) Find the value of $32^{3/5} \times 25^{1/2} \times 64^{-1/3}$.

32) Find the value of $4^{-1/2} + \left(\frac{1}{27}\right)^{1/3}$.

33) Find the values of $\left(\frac{1}{5}\right)^0$, $(125)^{-1/3}$, $(1\,000\,000)^{5/6}$.

34) Simplify: $(9x^4)^{-1/2}$, $(27x^6)^{-1/3}$, $(25a^8)^{-1/2}$.

35) Simplify: $(64a^6)^{1/2} \div (64a^6)^{-1/3}$.

THE EXPONENTIAL FUNCTION (e^x)

We can use the Binomial Theorem to derive a series for the important function denoted by e^x.

$$\left(1+\frac{x}{n}\right)^n = 1+n\left(\frac{x}{n}\right)+\frac{n(n-1)}{2!}\left(\frac{x}{n}\right)^2+\frac{n(n-1)(n-2)}{3\,!}\left(\frac{x}{n}\right)^3+\ldots$$

This series can be rewritten as

$$\left(1+\frac{x}{n}\right)^n = 1+\frac{x}{1!}+\frac{n^2\left(1-\frac{1}{n}\right)}{2!}\frac{x^2}{n^2}+\frac{n^3\left(1-\frac{1}{n}\right)\left(1-\frac{2}{n}\right)}{3!}\frac{x^3}{n^2}+\ldots$$

$$= 1+\frac{x}{1!}+\frac{1\left(1-\frac{1}{n}\right)}{2!}x^2+\frac{1\left(1-\frac{1}{n}\right)\left(1-\frac{2}{n}\right)}{3!}x^3+\ldots$$

As n tends to infinity terms like $\frac{1}{n}$, $\frac{2}{n}$ etc. will tend to zero. Hence

$$\lim_{n\to\infty}\left\{\left(1+\frac{x}{n}\right)^n\right\} = 1+\frac{x}{1!}+\frac{x^2}{2!}+\frac{x^3}{3!}+\ldots$$

$\lim\limits_{h\to\infty}\left\{\left(1+\frac{x}{n}\right)^n\right\}$ is given the symbol e^x and is called the **exponential function.**

It may be expressed as an infinite series, that is

$$e^x = 1+x+\frac{x^2}{2!}+\frac{x^3}{3!}+\frac{x^4}{4!}+\ldots$$

Now, if we differentiate this series term by term we obtain

$$\frac{d}{dx}(e^x) = 1+\frac{2x}{2!}+\frac{3x^2}{3!}+\frac{4x^3}{4!}+\ldots$$

and considering a typical coefficient, for example $\frac{4}{4!} = \frac{4}{4\times3\times2\times1} = \frac{1}{3!}$

then $$\frac{d}{dx}(e^x) = 1+x+\frac{x^2}{2!}+\frac{x^3}{3!}+\ldots$$

which is the same as the original series.

Hence $$\frac{d}{dx}(e^x) = e^x$$

This is the only mathematical function which does *not* change on differentiation.

Put another way we may say that for the graph of $y = e^x$ the rate of change $\frac{dy}{dx}$, at any point, is equal to e^x, that is $\frac{dy}{dx} = y$.

THE NUMERICAL VALUE OF e

Although we have defined the exponential function as e^x we shall not be able to make use of it unless we know the value of the constant e.

This may be found by substituting 1 for x in the series for e^x.

$$\text{Hence} \quad e = 1+1+\frac{1^2}{2!}+\frac{1^3}{3!}+\frac{1^4}{4!}+\frac{1^5}{5!}+\frac{1^6}{6!}+\frac{1^7}{7!}+\ldots$$

$$= 1+1+0.5+0.1667+0.0417+0.0083+0.0014+0.0002+\ldots$$

$$= 2.7183$$

$$\therefore \quad e = 2.718 \quad \text{correct to three decimal places.}$$

If greater accuracy is required more terms will have to be considered.

An alternative method of determining this value is to use continued fractions.

$$e = 2+\cfrac{1}{1+\cfrac{1}{2+\cfrac{1}{1+\ldots}}}$$

$$= 2+\frac{1}{1+}\,\frac{1}{2+}\,\frac{1}{1+}\,\frac{1}{1+}\,\frac{1}{4+}\,\frac{1}{1+}\,\frac{1}{1+}\,\frac{1}{6+}\,\frac{1}{1+}\,\frac{1}{1+}\,\frac{1}{8+}\,\frac{1}{1+}\ldots$$

Using a table to calculate the convergents, as shown in Chapter 5 of SCOTEC *Engineering Mathematics Level II*, we obtain:

	1	2	1	1	4	1	1	6	1	1	8	1
2	3	8	11	19	87	106	193	1264	1457	2721	23 225	25 946
1	1	3	4	7	32	39	71	465	536	1001	8 544	9 545

Again the convergents show that e = 2.718 correct to three decimal places.

THEORY OF LOGARITHMS

If N is a number such that

$$\boxed{N = a^x}$$

we say that x is the logarithm of N to the base a. We write

$$\boxed{\log_a N = x}$$

It should be carefully noted that

$$\text{number} = \text{base}^{\text{logarthim}}$$

Since $125 = 5^3$ we may write $\log_5 125 = 3$ and because $16 = 2^4$ we may write $\log_4 16 = 2$.

EXAMPLE 2

(a) If $\log_7 49 = x$, find the value x.

Writing the equation in index form we have

$$49 = 7^x$$
$$7^2 = 7^x$$

Hence $x = 2$ (since the indices on each side of the equation must be the same).

(b) If $\log_x 8 = 3$, find the value of x.

Writing the equation in index form we have

$$8 = x^3$$
$$2^3 = x^3$$

Hence $x = 2$ (since the indices on both sides of the equation are the same the bases must be the same).

Exercise 8

In each of the following find the value of x:

1) $\log_x 9 = 2$

2) $\log_x 81 = 4$

3) $\log_2 16 = x$

4) $\log_5 125 = x$

5) $\log_3 x = 2$

6) $\log_4 x = 3$

7) $\log_{10} x = 2$

8) $\log_7 x = 0$

9) $\log_x 8 = 3$

10) $\log_x 27 = 3$

RULES FOR THE USE OF LOGARITHMS

These rules are true for any chosen value of the base:

1) The logarithm of two numbers multiplied together may be found by adding their individual logarithms:

$$\log xy = \log x + \log y$$

2) The logarithm of two numbers divided may be found by subtracting their individual logarithms:

$$\log \frac{x}{y} = \log x - \log y$$

3) The logarithm of a number raised to a power may be found by multiplying the power by the logarithm of the number:

$$\log x^n = n \, . \log x$$

LOGARITHMS TO THE BASE 10

Logarithms to the base 10 are called common logarithms and are given as $\log_{10}$. When using logarithmic tables to solve numerical problems, the values given in the tables will be to the base 10. It is assumed that the reader is familiar with the use of common logarithms.

LOGARITHMS TO THE BASE 'e'

In higher mathematics all logarithms are taken to the base e, where e = 2.718 28. Logarithms to this base are often called natural logarithms. They are also called Napierian or hyperbolic logarithms.

To avoid confusion the base of a logarithm should always be stated, e.g. $\log_{10}$, but in practice this is often omitted if it is reasonably obvious which base is being used.

In the examples which follow the base will always be given.

Common logarithms are given as $\log_{10}$ (or lg)
and natural logarithms are given as $\log_e$ (or ln).

CHOICE OF BASE

When using logarithmic tables for numerical calculations common logarithms, i.e. $\log_{10}$, are preferred. This is because they are simpler to use in table form than natural logarithms.

If an electronic calculating machine of the scientific type (i.e., having keys which give trignometrical and logarithmic functions in addition to the usual +, −, × and ÷ etc.) is used then it is just as easy to use logarithms to the base e. Some machines have keys for both $\log_e$ and $\log_{10}$ but on the more limited models only $\log_e$ is given.

The natural logarithms is found by using the $\log_e$ (or ln) key and the natural antilogarithm is found by using the e^x key.

In the worked examples which follow in this chapter two alternative methods of solution are given — one using logarithmic tables and common logarithms, and the other using calculating machine and natural logarithms.

EXAMPLE 3

Evaluate $3.714^{2.87}$.

Let $$x = 3.714^{2.87}$$

and taking logarithms of both sides we have:

$$\log x = \log 3.714^{2.87}$$

$$\therefore \quad \log x = 2.87 \times \log 3.714$$

$$\therefore \quad x = \text{antilog}\,(2.87 \times \log 3.714)$$

The base of the logarithms has not yet been chosen — the sequence given will be true for any base value.

(a) Calculations using common logarithmic tables, i.e. $\log_{10}$

$$x = \text{antilog}_{10}(2.87 \times \log_{10} 3.714)$$
$$= \text{antilog}_{10}(2.87 \times 569\,9)$$
$$= \text{antilog}_{10}\, 1.636$$
$$= 43.25$$

Number	$\log_{10}$
2.87 0.569 9	0.457 9 $\bar{1}.755\,8$
1.636	0.213 7

or alternatively:

(b) Calculations using a machine with natural logarithms, i.e. $\log_e$, the sequence of operations being given on a flow chart (see opposite).

It should be noted that the answers obtained by the two methods are slightly different. This is because the tables used are only of the four figure type, and accuracy of the last figure cannot be guaranteed.

The sequence of operations would then be:

[3] [·] [7] [1] [4] [ln] [×] [2] [·] [8] [7] [=] [e^x]

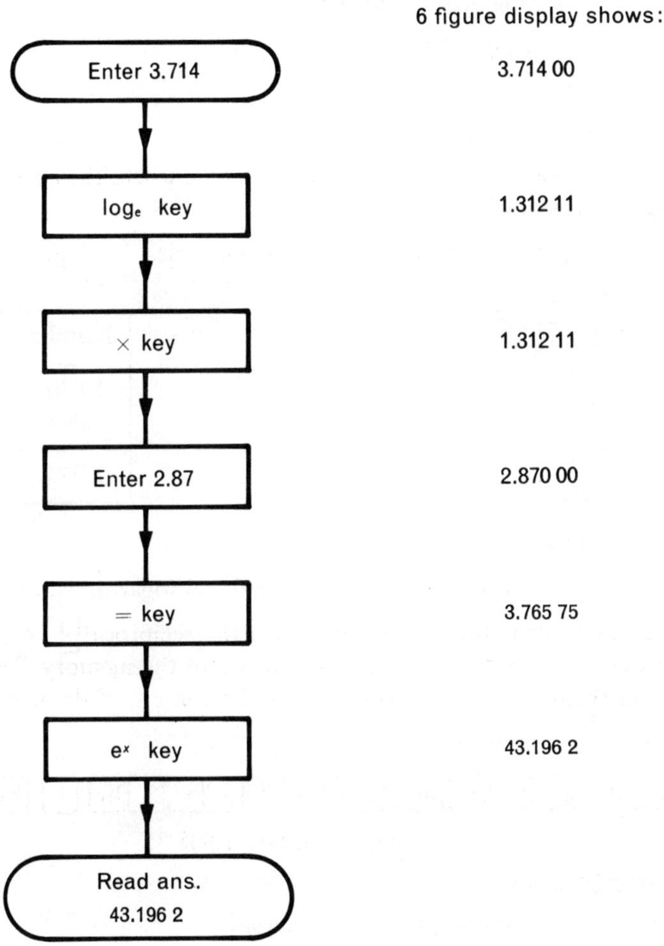

Fig. 3.1

INDICIAL EQUATIONS

These are equations in which the number to be found is an index, or part of an index.

The method of solution is to reduce the given equation to an equation involving logarithms, as the following examples will illustrate:

EXAMPLE 4

If $8.79^x = 67.65$ find the value of x.

Now taking logarithms of both sides of the given equation we have

$$\log 8.79^x = \log 67.65$$

$$\therefore \quad x \,.\, \log 8.79 = \log 67.65$$

$$\therefore \quad x = \frac{\log 67.65}{\log 8.79}$$

The base of the logarithms has not yet been chosen, the above procedure being true for any base value.

(a) Calculations using common logarithmic tables, i.e. $\log_{10}$

$$x = \frac{\log_{10} 67.65}{\log_{10} 8.79}$$

$$= \frac{1.830\,2}{0.944\,0}$$

$$= 1.938$$

Number	$\log_{10}$
1.830 2 0.944 0	0.262 5 $\bar{1}.975\,0$
1.938	0.287 5

or alternatively:

(b) Calculations using a machine with natural logarithms, i.e. $\log_e$

The quickest way, that is without using the reciprocal $\boxed{\frac{1}{x}}$ key, is to find the value of the bottom line and add this into the memory. Then find the value of the top line and divide this by the content of the memory.

This sequence would be

CM 8 · 7 9 ln M+ 6 7 · 6 5 ln ÷ MR =

giving an answer 1.938 86

EXAMPLE 5

Find the value of x if $1.793^{x+3} = 20^{0.982}$.

Now taking logarithms of both sides of the given equation we have

$$\log 1.793^{x+3} = \log 20^{0.982}$$

$$\therefore \quad (x+3)(\log 1.793) = (0.982)(\log 20)$$

$$\therefore \quad x+3 = \frac{(0.982)(\log 20)}{\log 1.793}$$

$$\therefore \quad x = \frac{(0.982)(\log 20)}{\log 1.793} - 3$$

The base of the logarithms has not yet been chosen, the above procedure being true for any base value.

(a) Calculations using common logarithic tables, i.e. $\log_{10}$

$$x = \frac{(0.982)(\log_{10} 20)}{\log_{10} 1.793} - 3$$
$$= \frac{(0.982)(1.301\,0)}{0.253\,6} - 3$$
$$= 5.032 - 3$$
$$= 2.032$$

Number	$\log_{10}$
0.982 1.301 0	$\bar{1}.991\,7$ 0.114 2
 0.253 6	0.105 9 $\bar{1}.404\,1$
5.032	0.701 8

or alternatively;

(b) Calculations using a machine with natural logarithms, i.e. $\log_e$:

The procedure will be similar to that used in Example 4. The sequence of operations is then

[CM] [1] [·] [7] [9] [3] [ln] [M+] [2] [0] [ln] [×] [0] [·] [9] [8] [2]
[=] [÷] [MR] [=] [−] [3] [=]

giving an answer 2.038 29

Exercise 9

Evaluate the following:

1) $\sqrt[3]{9.253}$

2) $\sqrt[4]{0.023\,8}$

3) $\sqrt{0.817\,6^3}$

4) $\sqrt[5]{7.891^3}$

5) $0.062\,9^{2/5}$

6) $0.371\,6^{7/3}$

7) $\sqrt{\dfrac{11.16 \times 0.329\,8}{7.315 \times 0.897\,4}}$

8) $\sqrt[4]{\dfrac{0.352 \times 17.26^2}{11.15 \times 0.93^3}}$

9) $[(0.816\,3)^2 \times (7.315)^4]^{1/5}$

10) $\dfrac{0.315\,7^2 \times \sqrt{18.16}}{\sqrt[3]{0.001\,76 \times 49.18}}$

11) $11.57^{0.3}$

12) $15.62^{2.15}$

13) $0.632\,7^{0.5}$

14) $0.065\,21^{3.16}$

15) $27.15^{-0.4}$

16) $73.2^{0.298}$

17) $0.5987^{-2.18}$

18) $3.761^{-1.79}$

19) $0.211^{-0.816}$

20) $\dfrac{5}{0.3168^{0.25}}$

21) $\dfrac{0.798^{0.3}}{18.16^{0.2}}$

22) $\dfrac{37.5}{0.25^{0.2}\times 0.03^{0.4}}$

23) $28.7^{0.62}\times 0.632$

24) $\dfrac{3.261^{0.3}}{\sqrt[6]{2.817}}$

Find the value of x in the following:

25) $3.6^x = 9.7$

26) $0.9^x = 2.176$

27) $\left(\dfrac{1}{7.2}\right)^x = 1.89$

28) $1.4^{(x+2)} = 9.3$

29) $21.9^{(3-x)} = 7.334$

30) $2.79^{(x-1)} = 4.377^x$

31) $\left(\dfrac{1}{0.64}\right)^{(2+x)} = 1.543^{(x+1)}$

32) $\dfrac{1}{0.9^{(x-2)}} = 8.45$

NATURAL LOGARITHMIC TABLES

In most books of mathematical tables there is a table of natural logarithms. Part of such a table is shown below:

Hyperbolic, Natural or Napierian Logarithms:

4.5	1.5041	5063	5085	5107	5129	5151	5173	5195	5217	5239	2	4	7	9	11	13	15	18	20
4.6	1.5261	5282	5304	5326	5347	5369	5390	5412	5433	5454	2	4	6	9	11	13	15	17	19
4.7	1.5476	5497	5518	5539	5560	5581	5602	5623	5644	5665	2	4	6	8	11	13	15	17	19
4.8	1.5686	5707	5728	5748	5769	5790	5810	5831	5851	5872	2	4	6	8	10	12	14	16	19
4.9	1.5892	5913	5933	5953	5974	5994	6014	6034	6054	6074	2	4	6	8	10	12	14	16	18
5.0	1.6094	6114	6134	6154	6174	6194	6214	6233	6253	6273	2	4	6	8	10	12	14	16	18
5.1	1.6292	6312	6332	6351	6371	6390	6409	6429	6448	6467	2	4	6	8	10	12	14	16	18
5.2	1.6487	6506	6525	6544	6563	6582	6601	6620	6639	6658	2	4	6	8	10	11	13	15	17
5.3	1.6677	6696	6714	6734	6752	6771	6790	6808	6827	6845	2	4	6	7	9	11	13	15	17
5.4	1.6864	6882	6901	6919	6938	6956	6974	6993	7011	7029	2	4	5	7	9	11	13	15	17

Natural logarithms of 10^{+n}:

n	**1**	**2**	**3**	**4**	**5**	**6**	**7**	**8**	**9**
$\log_e 10^n$	2.3026	4.6052	6.9078	9.2103	11.5129	13.8155	16.1181	18.4207	20.7233

Natural logarithms of 10^{-n}:

n	**1**	**2**	**3**	**4**	**5**	**6**	**7**	**8**	**9**
$\log_e 10^{-n}$	$\bar{3}.6974$	$\bar{5}.3948$	$\bar{7}.0922$	$\overline{10}.7897$	$\overline{12}.4871$	$\overline{14}.1845$	$\overline{17}.8819$	$\overline{19}.5793$	$\overline{21}.2767$

The first column of a set of full tables gives the natural logarithms of numbers from 1.0 to 9.9 (the specimen table above only gives numbers from 4.5 to 5.49), and the tables are read in the same way as ordinary log tables except that the characteristic is also given. Thus:

$$\log_e 4.568 = 1.519\,1$$

When a natural logarithm of a number, which lies outside the tabulated range, is required the subsidiary table has to be used. The following examples show how this is done.

EXAMPLE 6

To find $\log_e 483.4$.

$$483.4 = 4.834 \times 100 = 4.834 \times 10^2$$

$$\therefore \quad \log_e 483.4 = \log_e 4.834 + \log_e 10^2$$

From the main table: $\log_e 4.834 = 1.575\,6.$

From the subsidiary table: $\log_e 10^2 = 4.605\,2.$

$$\therefore \quad \log_e 483.4 = 1.575\,6 + 4.605\,2 = 6.180\,8$$

EXAMPLE 7

To find $\log_e 0.053\,61$.

$$0.053\,61 = \frac{5.361}{100} = \frac{5.361}{10^2} = 5.361 \times 10^{-2}$$

$$\therefore \quad \log_e 0.053\,61 = \log_e 5.361 + \log_e 10^{-2}$$

From the main table: $\log_e 5.361 = 1.679\,2$

From the subsidiary table: $\log_e 10^{-2} = \bar{5}.394\,8$

$$\therefore \quad \log_e 0.053\,61 = 1.679\,2 + \bar{5}.394\,8 = \bar{3}.074\,0$$

$$= -3 + 0.074\,0 = -2.926\,0$$

CONVERSION OF NATURAL LOGARITHMS TO COMMON LOGARITHMS

The use of tables of natural logarithms is rather tedious as the last two worked examples show. Their use may be avoided by finding the common logarithm and then converting to a natural logarithmic value.

Suppose we have a number, N, and wish to find the value of its natural logarithm, that is $\log_e N$.

Let $\quad x = \log_e N$

then $\quad e^x = N$

and taking common logarithms of both sides we have

$$\log_{10} e^x = \log_{10} N$$

$$\therefore \quad x \cdot \log_{10} e = \log_{10} N$$

$$\therefore \quad x = \frac{\log_{10} N}{\log_{10} e}$$

or

$$\log_e N = \frac{\log_{10} N}{\log_{10} e}$$

The value of $\log_{10} e = \log_{10} 2.718\,3 = 0.434\,3$

hence:

$$\log_e N = \frac{\log_{10} N}{0.434\,3}$$

or:

$$\log_e N = 2.3026 \times \log_{10} N$$

Hence we may find the natural logarithm of a number by first finding its common logarithm and then multiplying by 2.302 6 (or dividing by 0.434 3).

EXAMPLE 8

By using common logarithms find the value of $\log_e 25$.

We have:

$$\log_e N = 2.302\,6 \times \log_{10} N$$

$$\therefore \quad \log_e 25 = 2.302\,6 \times \log_{10} 25$$

$$= 2.302\,6 \times 1.397\,9$$

$$= 3.218\,9$$

EXAMPLE 9

By using common logarithms find $\log_e 0.396\,4$.

We have:

$$\log_e 0.396\,4 = 2.302\,6 \times \log_{10} 0.396\,4$$

$$= 2.302\,6 \times \bar{1}.598\,1$$

$$= 2.302\,6(-1 + 0.598\,1)$$

$$= 2.302\,6(-0.401\,9)$$

$$= -0.925\,4$$

EXAMPLE 10

Find the number whose natural logarithm is 0.631.

If we let the required number be N then:

$$\log_e N = 0.631$$

but we know that: $\log_e N = 2.302\,6 \times \log_{10} N$

$$\therefore \quad 0.631 = 2.302\,6 \times \log_{10} N$$

or: $$\log_{10} N = \frac{0.631}{2.302\,6}$$

$$= 0.274\,0$$

$$\therefore \quad N = \text{antilog}_{10} 0.274\,0$$

$$= 1.879$$

Problems involving the use of natural logarithms occur frequently in engineering.

Exercise 10

1) Use tables of natural logarithms to find the values of:

(a) $\log_e 311.3$ (b) $\log_e 7.081$ (c) $\log_e 20.15$
(d) $\log_e 8.128$ (e) $\log_e 0.491\,3$ (f) $\log_e 0.008\,178$

2) By using common logarithms find the values of:

(a) $\log_e 36$ (b) $\log_e 482$ (c) $\log_e 0.955\,4$
(d) $\log_e 0.038\,1$ (e) $\log_e 2.07$ (f) $\log_e 0.10$

VALUES OF e^x AND e^{-x}

Most mathematical tables which include logarithms and trignometrical functions also include a table giving values of e^x and e^{-x} for values of x from 0 to 6. It may appear that the range of x values is rather limited but it is adequate for most practical problems as may be seen from the examples which follow later in this chapter.

EXPONENTIAL GRAPHS

Curves which have equations of the type e^x and e^{-x} are called exponential graphs.

We may plot the graphs of e^x and e^{-x} by using mathematical tables to find values of e^x and e^{-x} for chosen values of x. We should remember that any number to a zero power is unity: hence $e^0 = 1$.

Drawing up a table of values we have:

x	-2	-1	0	1	2
e^x	0.14	0.37	1	2.72	7.39
$-x$	2	1	0	-1	-2
e^{-x}	7.39	2.72	1	0.37	0.14

For convenience both the curves are shown plotted on the same axes in Fig. 3.2. Although the range of values chosen for x is limited, the overall shape of the curves is clearly shown.

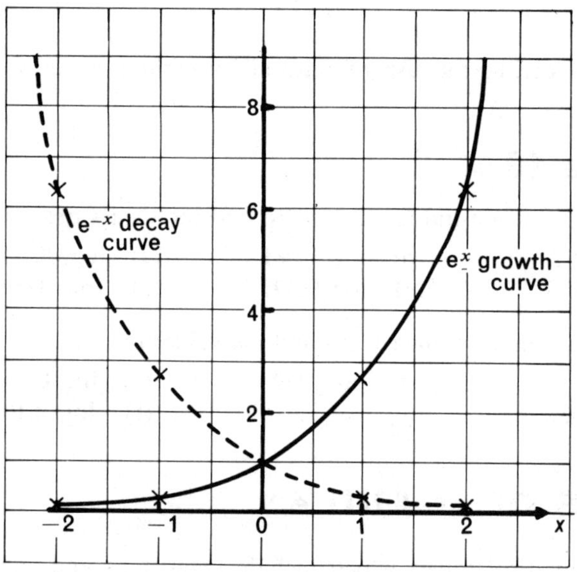

Fig. 3.2

The rate at which a curve is changing at any point is given by the gradient of the tangent at that point.

Remember the sign convention for gradients is:

positive gradient   negative gradient

Now the gradient at any point on the e^x graph is positive and so the rate of change is positive. In addition the rate of change increases as the values of x increase. A graph of this type is called a growth curve.

The gradient at any point on the e^{-x} graph is negative and so the rate of change is negative. In addition the rate of change decreases as the values of x decrease. A graph of this type is called a decay curve.

EXAMPLE 11

The instantaneous e.m.f. in an inductive circuit is given by the expression $100.e^{-4t}$ volts, where t is time in seconds. Plot the graph of the e.m.f. for values of t from 0 to 0.5 seconds, and use the graph to find:

(a) the value of the e.m.f. when $t = 0.25$ seconds, and

(b) the rate of change of the e.m.f. when $t = 0.1$ seconds.

Check this result using the method of differentiation.

First we will draw up a table of values taking values of t at 0.1 s intervals:

t	0	0.1	0.2	0.3	0.4	0.5
$-4t$	0	−0.4	−0.8	−1.2	−1.6	−2.0
e^{-4t}	1	0.67	0.45	0.30	0.20	0.14
$100e^{-4t}$	100	67	45	30	20	14

The graph is shown plotted in Fig. 3.3.

(a) The point P on the curve is at 0.25 seconds shown on the t scale and the corresponding value of e.m.f. can be read directly from the vertical axis scale. The value is 37 volts.

(b) The point Q on the graph is at 0.1 seconds. Now the rate of change of the curve at Q is given by the gradient of the tangent at Q. This gradient may be found by constructing a suitable right-angled triangle such as MNO in Fig. 3.3, and finding the ratio $\frac{MO}{ON}$.

Hence the gradient at Q

$$= \frac{MO}{ON} = \frac{92 \text{ volts}}{0.34 \text{ seconds}} = 270 \text{ volts per second.}$$

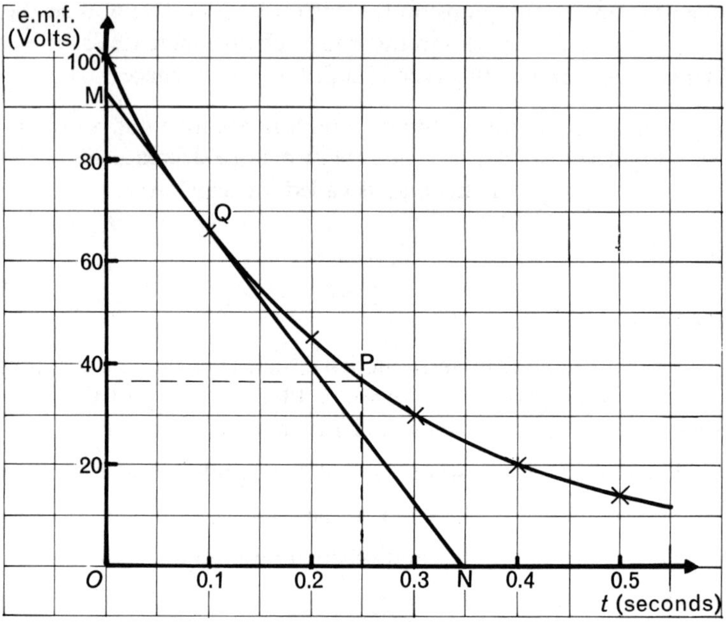

Fig. 3.3

According to the sign convention a line sloping downwards from left to right has a negative gradient.

Hence the gradient at Q is -270 volts per second, which means that the rate of change of the curve at Q is -270 volts per second.

This is the same as saying that the e.m.f. at $t = 0.1$ seconds is decreasing at the rate of 270 volts per second.

A Check Using Differentiation

The rate of change of the e.m.f. at any instant is given by the gradient of the e.m.f.–time curve and using the calculus this is denoted by $\frac{\mathrm{d}}{\mathrm{d}t}(\text{e.m.f.})$.

Since we have $\quad \text{e.m.f.} = 100.\mathrm{e}^{-4t}$

then $\quad \frac{\mathrm{d}}{\mathrm{d}t}(\text{e.m.f.}) = 100(-4)\mathrm{e}^{-4t}$

$$= -400\mathrm{e}^{-4t}$$

Hence when $t = 0.1$ seconds

then the required rate of change $= -400.e^{-4(0.1)}$

$= -400.e^{-0.4}$

$= -268$ volts/second

This verifies the approximate value of -270 volts/second obtained by use of the graph.

EXAMPLE 12

The formula $i = 2(1-e^{-10t})$ gives the relationship between the instantaneous current i amperes and the time t seconds in an inductive circuit. Plot a graph of i against t, taking values of t from 0 to 0.3 seconds at intervals of 0.05 seconds. Hence find:

(a) the initial rate of growth of the current i when $t = 0$, and

(b) the time taken for the current to increase from 1 to 1.6 amperes.

Verify these results using theoretical methods.

The table of values is drawn up as follows:

t	0	0.05	0.10	0.15	0.20	0.25	0.30
$-10t$	0	-0.5	-1.0	-1.5	-2.0	-2.5	-3.0
1	1	1	1	1	1	1	1
e^{-10t}	1	0.61	0.37	0.22	0.14	0.08	0.05
$(1-e^{-10t})$	0	0.39	0.63	0.78	0.86	0.92	0.95
$2(1-e^{-10t})$	0	0.78	1.26	1.56	1.72	1.84	1.90

The curve is shown plotted in Fig. 3.4.

(a) When $t = 0$ the initial rate of growth will be given by the gradient of the tangent at O. The tangent at O is the line OM and its gradient may be found by using a suitable right angled triangle MNO and finding the ratio $\frac{MN}{ON}$.

Hence the initial rate of growth of $i = \frac{MN}{ON} = \frac{2 \text{ amperes}}{0.1 \text{ seconds}}$

$= 20$ amperes per second.

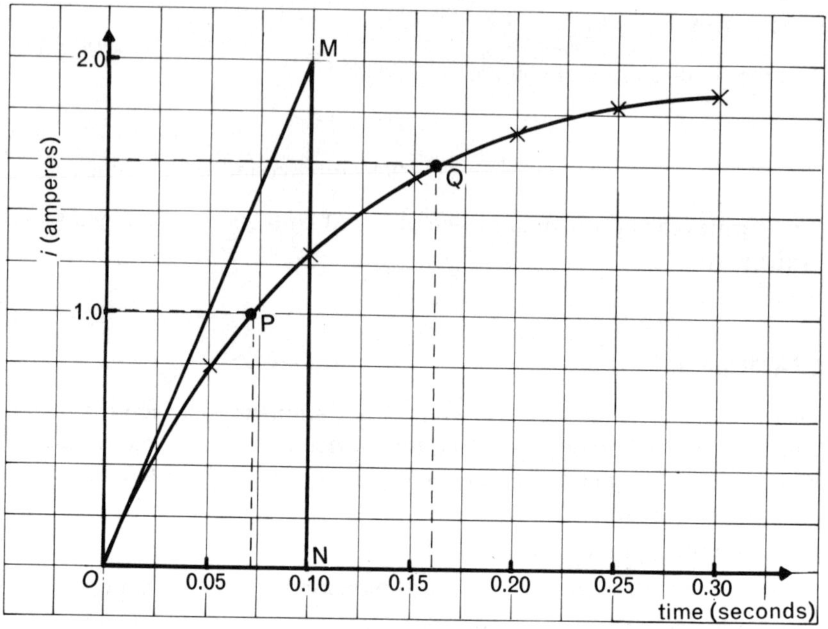

Fig. 3.4

(b) The point P on the curve corresponds to a current of 1.0 amperes and the time at which this occurs may be read from the t scale and is 0.07 seconds.

Similarly point Q corresponds to a current of 1.6 ampere and occurs at 0.16 seconds.

Hence the time between P and Q is $0.16 - 0.07 = 0.09$ seconds.

This means that the time for the current to increase from 1 to 1.6 amperes is 0.09 seconds.

Solution using Theoretical Methods

(i) The rate of growth of current is another way of stating a positive rate of change of current. At any instant this is given by the gradient of the current-time curve and using the calculus this is denoted by $\frac{d}{dt}$ (current).

Since we have $\quad \text{current} = 2(1 - e^{-10t})$

or $\quad i = 2 - 2.e^{-10t}$

then $\quad \frac{di}{dt} = -2(-10)e^{-10t} = 20e^{-10t}$

Hence when $t = 0$ seconds

$$\begin{aligned}\text{the rate of change of current} &= 20.\mathrm{e}^{-10(0)} \\ &= 20.\mathrm{e}^{0} \\ &= 20 \text{ A/s} \qquad (\text{since } \mathrm{e}^0 = 1)\end{aligned}$$

This verifies the value obtained by use of the graph.

(ii) We have $i = 2(1-\mathrm{e}^{-10t})$

hence when $i = 1.6$ A then

$$1.6 = 2(1-\mathrm{e}^{-10t})$$

and rearranging $\mathrm{e}^{-10t} = 1-\dfrac{1.6}{2} = 0.2$

and rewriting this equation in logarithmic form

$$-10t = \log_e 0.2$$

$$\therefore \quad t = -\frac{1}{10}\log_e 0.2 = 0.161 \text{ s}$$

Also when $i = 1.0$ A then

$$1 = 2(1-\mathrm{e}^{-10t})$$

and rearranging $\mathrm{e}^{-10t} = 1-\dfrac{1}{2} = 0.5$

and rewriting this equation in logarithmic form

$$-10t = \log_e 0.5$$

$$\therefore \quad t = -\frac{1}{10}\log_e 0.5 = 0.069 \text{ s}$$

Now

required time = (time to increase to 1.6 A)−(time to increase to 1.0 A)

$$= 0.161-0.069 = 0.092 \text{ s}$$

This verifies the more approximate value obtained by use of the graph.

Exercise 11

1) Find the values of:

(a) $\log_e 9.137$ (b) $\log_e 6.017$

(c) $\log_e 29.16$ (d) $\log_e 298.2$

(e) $\log_e 0.5917$ (f) $\log_e 0.007\,138$

2) A formula used for calculating the self-inductance of parallel conductors is:

$$L = 0.000\,644\left\{\log_e \frac{d}{r} + 0.25\right\}$$

Find L when $d = 50$ and $r = 0.45$.

3) The formula:

$$L = 0.002 \log_e \frac{4l}{d} - 0.002$$

is used in connection with straight aerials. Calculate the value of L when $l = 250$ and $d = 0.25$.

4) Find the values of:

(a) $e^{0.3}$ (b) $e^{2.5}$ (c) $e^{4.5}$

(d) $e^{-0.4}$ (e) e^{-1} (f) $e^{-3.5}$

5) Plot a graph of $y = e^{2x}$ for values of x from -1 to $+1$ at 0.25 unit intervals. Use the graph to find the value of y when $x = 0.3$, and the value of x when $y = 5.4$.

6) Using values of x from -4 to $+4$ at one unit intervals plot a graph of $y = e^{-x/2}$. Hence find the value of x when $y = 2$, and the gradient of the curve when $x = 0$.

7) For a constant pressure process on a certain gas the formula connecting the absolute temperature T and the specific entropy s is $T = 24.e^{3s}$. Plot a graph of T against s taking values of s equal to 1.000, 1.033, 1.066, 1.100, 1.133, 1.166 and 1.200. Use the graph to find the value of:

(a) T when $s = 1.09$.

(b) s when $T = 700$.

8) The equation $i = 2.4e^{-6t}$ gives the relationship between the instantaneous current, i mA, and the time, t seconds. Plot a graph of i against t for values of t from 0 to 0.6 seconds at 0.1 second intervals. Use the curve obtained to find the rate at which the curent is decreasing when $t = 0.2$ seconds. (The more able student may wish to check the results obtained graphically by a theoretical method using calculus.)

9) In a capacitive circuit the voltage v and the time t seconds are connected by the relationship $v = 240(1 - e^{-5t})$. Draw the curve of v against t for values of $t = 0$ to $t = 0.7$ seconds at 0.1 second intervals. Hence find:

(a) the time when the voltage is 140 volts, and

(b) the initial rate of growth of the voltage when $t = 0$.
(The more able student may wish to check the results obtained graphically by a theoretical method using calculus.)

10) A radioactive material decays according to the law $N = N_i(e^{-0.0693t})$ where N is the amount of radioactivity remaining, N_i is the initial amount of radioactivity, and t years is the time taken to decay from N_i to N. Plot a graph of N (this will be in terms of N_i, e.g., $0.4N_i$) against t for values of $t = 0$ to $t = 30$ years at 5 yearly intervals. Hence find the half life of the material (that is when half the radioactivity has decayed or $N = \frac{1}{2}N_i$).

PROJECTS

1) The exponential function initiates ideas for projects in various topics. For example, decay of current along a wire; the disintegration of a radioactive element; atmospheric pressure; the cooling of hot metal; compound interest; the growth of bacilli; the absorption of gamma rays; damped oscillations; etc.

2) Paper can be ruled in a great variety of specialised ways. In the next chapter we study logarithmic graph paper.

List the type and purpose of some rulings which may be used in the different branches of your industry.

4. LOGARITHMIC GRAPH PAPER

On reaching the end of this chapter you should be able to:-

1. *Plot experimental data on log–log and log–linear graph paper.*
2. *Determine laws of experimental data of the form* $z = at^n$, $z = a.b^t$, *and* $z = a.e^{bt}$.

NON-LINEAR LAWS WHICH CAN BE REDUCED TO THE LINEAR FORM

Many non-linear equations can be reduced to the linear (i.e., straight line) form by making a suitable substitution. We have met previously equations such as $y = m.t^2 + c$ and we put $x = t^2$ so that the equation becomes $y = mx + c$ which is a straight line when plotted and enables us to find values of the constants m and c.

This section is similar except that logarithmic and exponential relationship are covered.

THE STRAIGHT LINE

You will remember the standard equation of the straight line is $y = m.x + c$ where m and c are constants.

The values of these constants may be found using information obtained from either Fig. 4.1 or Fig. 4.2.

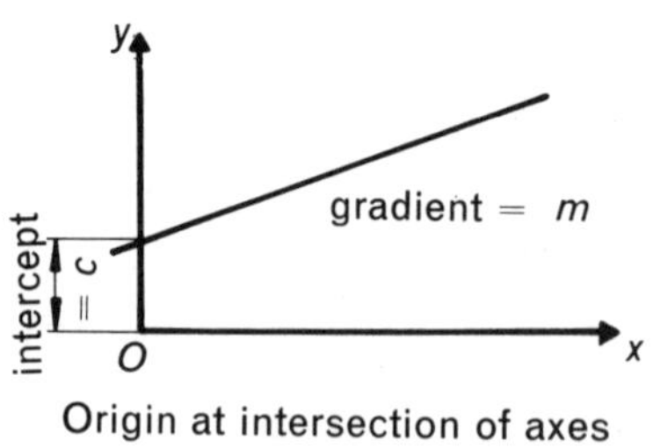

Fig. 4.1

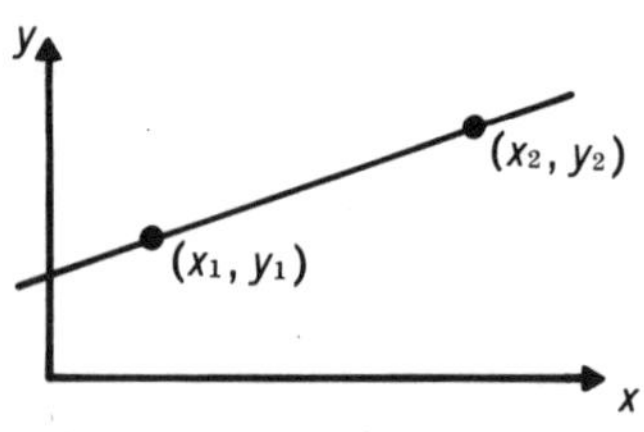

Fig. 4.2

If the origin is at the intersection of the axes the values of m and c may be found using Fig. 4.1.

If the origin is not shown the two point method must be used (Fig. 4.2), i.e., since the point (x_1, y_1) lies on the line then $y_1 = mx_1+c$ and the point (x_2, y_2) lies on the line then $y_2 = mx_2+c$.

These equations may be solved simultaneously to find the values of m and c.

The two point method may be used whether or not the origin is shown and should be used in all the work which follows in this chapter.

EXAMPLE 1

The table below gives corresponding values of x and y. Plot these values on a suitable graph and hence find the equation connecting x and y.

x	3	5	8	12
y	15	23	35	51

The graph is shown plotted in Fig. 4.3.

Since the plotted points lie on a straight line the equation connecting x and y will be of the form $y = mx+c$.

To find the values of the constants m and c choose two points P and Q which lie on the line. The co-ordinates of P and Q will satisfy $y = mx+c$.

Hence at the point Q (10, 43) then $43 = m(10)+c$ [1]

and at the point P (4, 19) then $19 = m(4)+c$ [2]

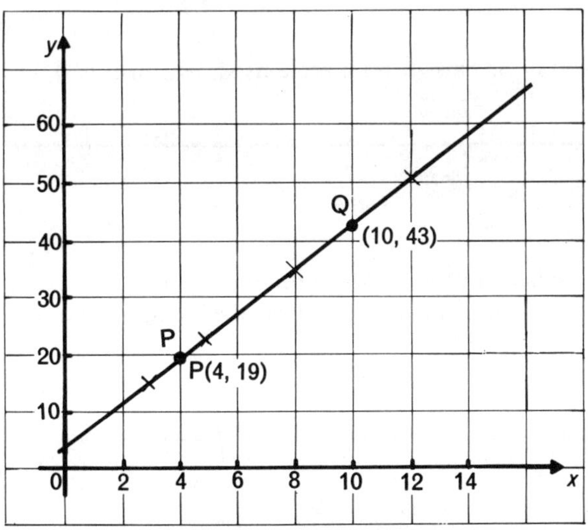

Fig. 4.3

Subtracting equation [2] from [1], $24 = 6m$

$\therefore$ $m = 4$

and substituting $m = 4$ into equation [1]

then $43 = 4(10)+c$

$\therefore$ $c = 3$

Hence the required equation is $y = 4x+3$.

LOGARITHMIC SCALES

It has been shown earlier that: Number = $(\text{base})^{\text{logarthim}}$

and if we use a base of 10 then: Number = $(10)^{\text{logarithm}}$

Since $1000 = 10^3$ then we may write $\log_{10}1000 = 3$

and since $100 = 10^2$ then we may write $\log_{10}100 = 2$

and since $10 = 10^1$ then we may write $\log_{10}10 = 1$

and since $1 = 10^0$ then we may write $\log_{10}1 = 0$

and since $0.1 = 10^{-1}$ then we may write $\log_{10}0.1 = -1$

and since $0.01 = 10^{-2}$ then we may write $\log_{10}0.01 = -2$

and since $0.001 = 10^{-3}$ then we may write $\log_{10}0.001 = -3$ and so on.

These logarithms may be shown on a scale as shown in Fig. 4.4.

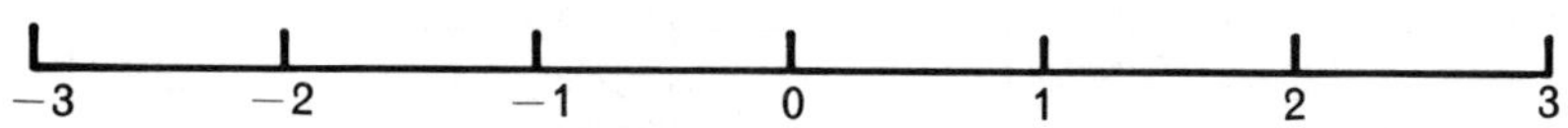

Fig. 4.4

However since we wish to plot numbers directly on to the scale (without any reference to their logarithms) the scale is labelled as shown in Fig. 4.5.

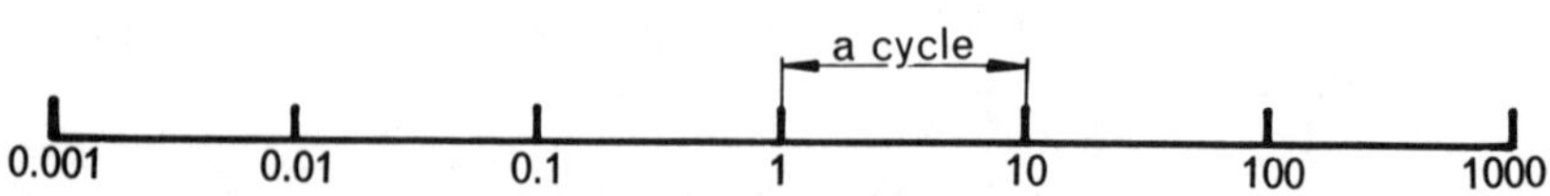

Fig. 4.5

Each division is called a cycle and is sub-divided using a logarithmic scale as, for instance, the scales on a slide rule. Two such cycles are shown in Fig. 4.6.

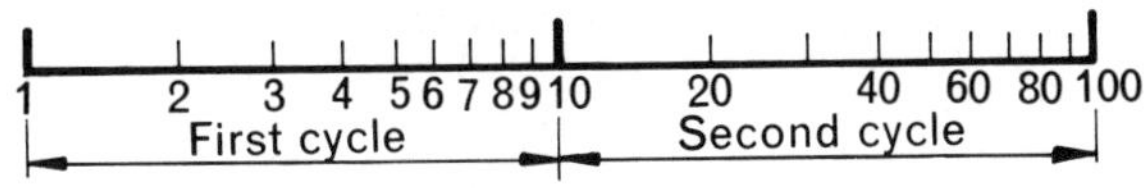

Fig. 4.6

The choice of numbers on the scale depends on the numbers allocated to the variable in the problem to be solved. Thus in Fig. 4.6 the numbers run from 1 to 100.

In all the work which follows in this chapter the logarithms used will be to the base 10. The 10 is often omitted and so, for instance, $\log x$ means $\log_{10} x$.

Consider the following relationships in which z and t are the variables, whilst a, b, and n are constants.

$z = a.t^n$

Now $$z = a.t^n$$

and taking logs $$\log z = \log (a.t^n)$$

$$= \log t^n + \log a$$

$\therefore$ $$\log z = n.\log t + \log a$$

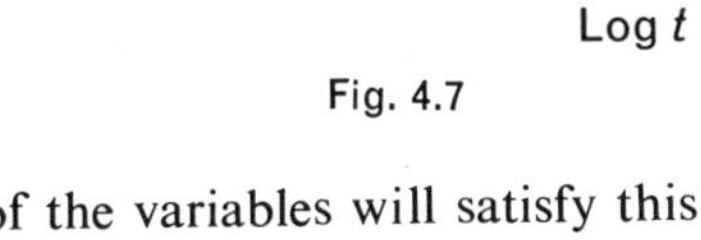

Fig. 4.7

The given values of the variables will satisfy this equation if they satisfy the original equation. Comparing this equation with $y = mx + c$, which is the standard equation of a straight line, we see that if we plot $\log z$ on the y-axis and $\log t$ on the x-axis the result will be a straight line (Fig. 4.7) and the values of the constants n and a may be found using the two point method.

$z = a.b^t$

Now $$z = a.b^t$$

and taking logs $$\log z = \log (a.b^t)$$
$$= \log b^t + \log a$$

$\therefore$ $$\log z = (\log b)t + \log a$$

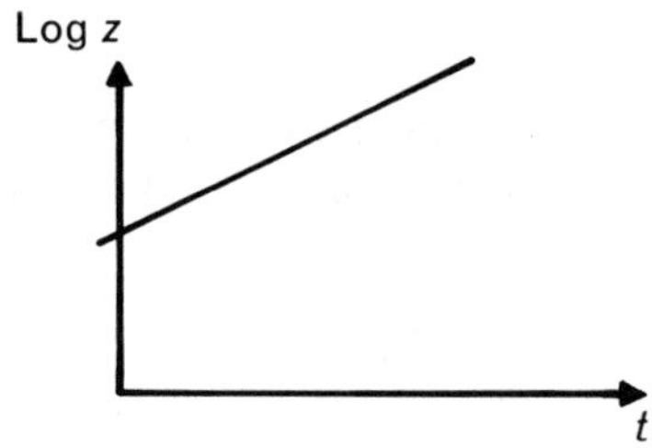

Fig. 4.8

We now proceed in a manner similar to that used for the previous equation by plotting $\log z$ on the y-axis and t on the x-axis, and again obtain a straight line (Fig. 4.8).

$z = a.e^{bt}$

Now $$z = a.e^{bt}$$

and taking logs $$\log z = \log (a.e^{bt})$$
$$= \log e^{bt} + \log a$$
$$= (b.\log e)t + \log a$$

but $\log e = 0.4343$

$\therefore$ $$\log z = (0.4343b)t + \log a$$

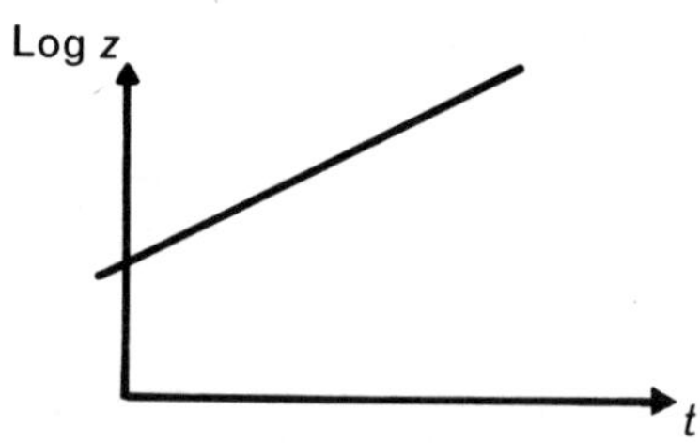

Fig. 4.9

Again proceeding in a manner similar to that used for the previous equations, we plot $\log z$ on the y-axis and t on the x-axis and again obtain a straight line (Fig. 4.9).

LOGARITHMIC GRAPH PAPER

Logarithmic scales (such as those used on a slide rule) may be used on graph paper in place of the more usual linear scales. By using graph paper ruled in this way log plots may be made without the necessity of looking up the logs of each given value. Semi-logarithmic graph paper is also available and has one way ruled with log scales whilst the other way has the usual linear scale. Examples of each are shown in Figs. 4.10 and 4.11.

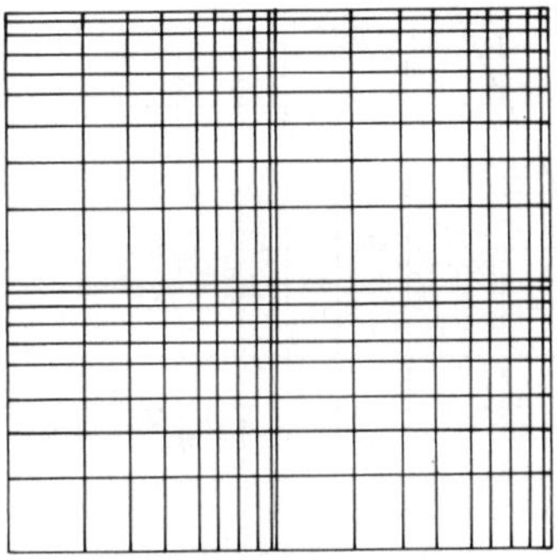

Full logarithmic rulings

Fig. 4.10

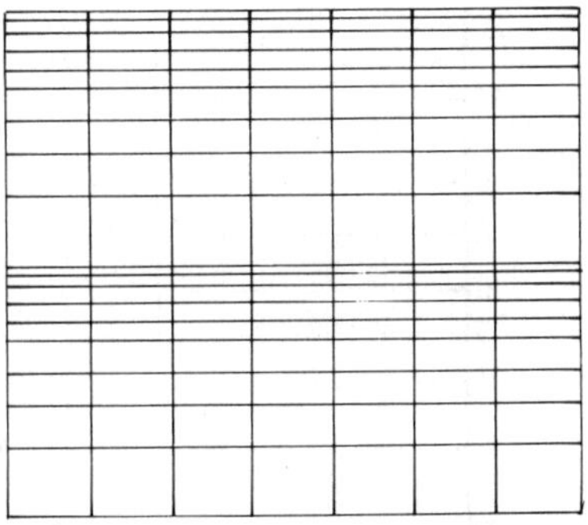

Semi-logarithmic rulings

Fig. 4.11

The use of logarithmic graph paper enables power or exponential relationships between two variables to be verified quickly.

A straight line graph on full logarithmic (or log–log) graph paper indicates a relationship between the variables z and t of the form $z = a.t^n$, a and n being constants.

A straight line graph on semi-logarithmic (or log–linear) graph paper indicates a relationship between the variables z and t of the form $z = a.b^t$ or $z = a.\mathrm{e}^{bt}$, a and b being constants.

The use of these scales and the special graph paper is shown by the examples which follow.

EXAMPLE 2

The law connecting two quantities z and t is of the form $z = a.t^n$. Find the law given the following pairs of values:

z	3.170	4.603	7.499	10.50	15.17
t	7.980	9.863	13.03	15.81	19.50

The relationship $z = a.t^n$

gives (see text) $\log z = n.\log t + \log a$ [1]

Hence we plot the given values of z and t on log scales as shown in Fig. 4.12.

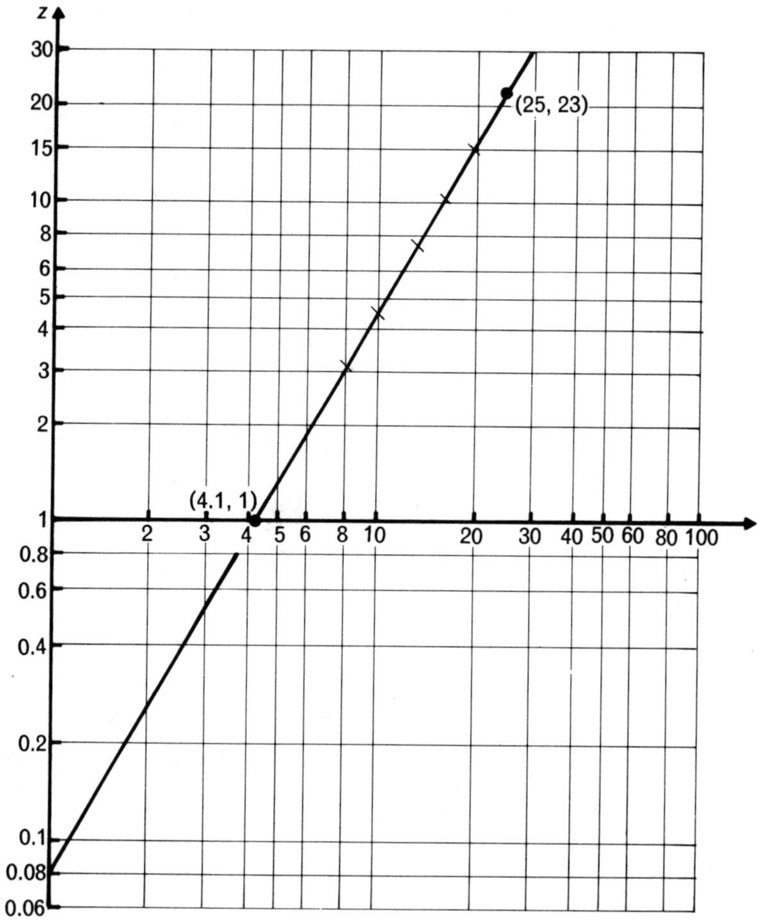

Fig. 4.12

On both the vertical and horizontal axes we require 2 cycles, the first for values from 1 to 10 and the second for values 10 to 100.

The constants are found by taking two pairs of coordinates:

Point (25, 23) lies on the line and putting these values in equation [1],

$$\log 23 = n\{\log 25\} + \log a \qquad [2]$$

Point (4.1, 1) lies on the line and putting these values in equation [1],

$$\log 1 = n\{\log 4.1\} + \log a \qquad [3]$$

Subtracting equation [3] from equation [2],

$$\log 23 - \log 1 = n\{\log 25 - \log 4.1\}$$

$$\log (23/1) = n\{\log (25/4.1)\}$$

$$n = \frac{\log (23/1)}{\log (25/4.1)} = \frac{\log 23}{\log 6.1} = \frac{1.367}{0.7853}$$

$$n = 1.74$$

Substituting this value of n in equation [3],

$$\log 1 = 1.74\{\log 4.1\} + \log a$$

$\therefore$
$$\log a = \log 1 - 1.74\{\log 4.1\}$$

$$= 0 - 1.075$$

$$= -1.075 = -1 - 1 + 1 - 0.075$$

$$= -2 + 0.925 = \bar{2}.925$$

$\therefore$
$$a = 0.084$$

Hence the law is $z = 0.084t^{1.74}$.

EXAMPLE 3

The table gives values obtained in an experiment. It is thought that the law may be of the form $z = a.b^t$, where a and b are constants. Verify this and find the law.

t	0.190	0.250	0.300	0.400
z	11 220	18 620	26 920	61 660

We think that the relationship is of the form:

$$z = a.b^t$$

which gives (see text),

$$\log z = (\log b)t + \log a \qquad [1]$$

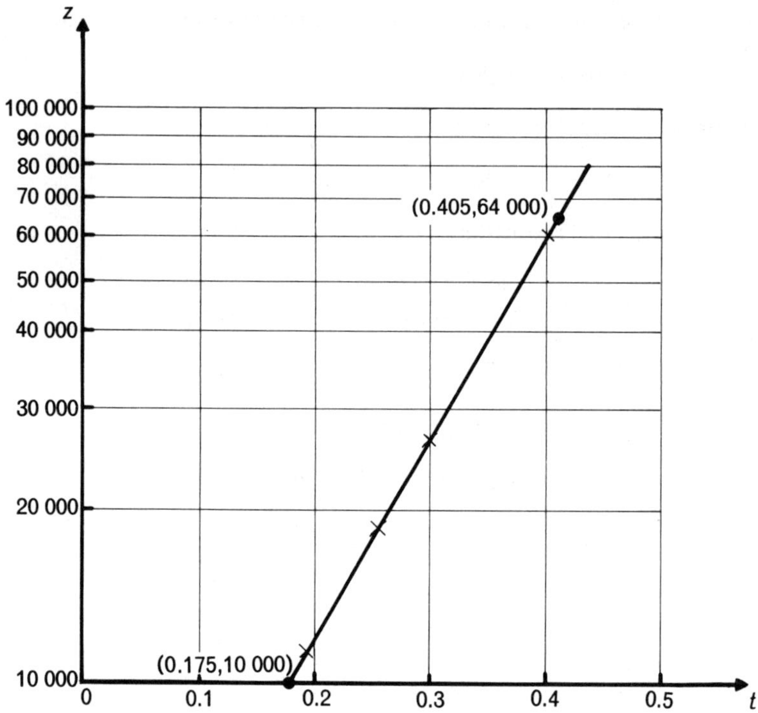

Fig. 4.13

Hence we plot the given values of z on a vertical log scale — the t values however will be on the horizontal axis on an ordinary linear scale (Fig. 4.13).

The points lie on a straight line, and as before the given values of z and t obey the law. We now have to find the co-ordinates of two points lying on the line.

Point (0.405, 64 000) lies on the line, and substituting in equation [1],

$$\log 64\,000 = (\log b)0.405+\log a \qquad [2]$$

Point (0.175, 10 000) lies on the line, and substituting in equation [1],

$$\log 10\,000 = (\log b)0.175+\log a \qquad [3]$$

Subtracting equation [3] from equation [2],

$$\log 64\,000-\log 10\,000 = (\log b)(0.405-0.175)$$

$$\therefore \qquad 4.8062-4.0000 = 0.230\ (\log b)$$

$$\therefore \qquad \log b = \frac{0.8062}{0.230} = 3.5052$$

$$\therefore \qquad b = 3200$$

Substituting in the equation [3],

$$\log 10\,000 = (3.5052)0.175 + \log a$$

$$\therefore \qquad \log a = \log 10\,000 - 3.5(0.175)$$

$$= 4 - 0.6134 = 3.3866$$

$$\therefore \qquad a = 2435$$

Hence the law is:

$$z = 2\,435(3200)^t$$

EXAMPLE 4

V and t are connected by the law $V = a.e^{bt}$. If the values given in the table satisfy the law, find the constants a and b.

t	0.05	0.95	2.05	2.95
V	20.70	24.49	30.27	36.06

The law is:

$$V = a.e^{bt}$$

which gives (see text),

$$\log V = 0.4343bt + \log a \qquad [1]$$

As in last example V values are plotted on a log scale on the vertical axis, whilst the t values are plotted on the horizontal axis on an ordinary linear scale (Fig. 4.14).

Point (3.15, 37.2) lies on the line, and substituting in equation [1],

$$\log 37.2 = (0.4343)b(3.15) + \log a \qquad [2]$$

Point (0.30, 21.5) lies on the line, and substituting in equation [1],

$$\log 21.5 = (0.4343)b(0.3) + \log a \qquad [3]$$

Subtracting,

$$\log 37.2 - \log 21.5 = (0.4343)b(3.15 - 0.3)$$

$$\therefore \qquad b = \frac{\log (37.2/21.5)}{(0.4343)(2.85)} = \frac{0.238}{(0.4343)(2.85)}$$

$$= 0.192$$

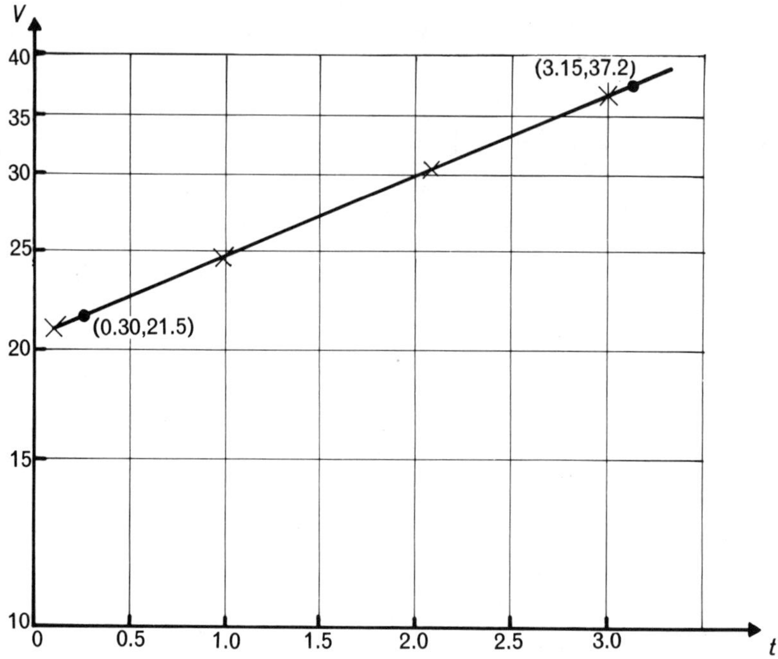

Fig. 4.14

Substituting in the equation [3]

$$\log 21.5 = (0.4343)(0.192)(0.3) + \log a$$

$$\therefore \quad \log a = 1.3324 - 0.0248 = 1.3076$$

$$\therefore \quad a = 20.3$$

Exercise 12

1) The resistance R of the field coils of a motor is measured at different temperatures and found to be as follows:

t (°C)	20	30	40	50	60	70	80
R (ohm)	101	104.6	108.1	111.6	115.1	118.5	122

Plot the graph of R against t. If R and t are related by a law of the form $R = a+bt$, find from the graph the values of the constants a and b.

2) In a test on a Lechlanché cell the terminal voltage (V) was measured for various values of current (I) drawn from the cell. The results obtained are shown below.

V	1.44	1.32	1.22	1.06	0.96	0.84
I	0.1	0.3	0.5	0.7	0.9	1.1

Show that the law connecting V and I is of the form $V = E - Ir$ and find values for the constants E (the e.m.f. of the cell) and r (the internal resistance of the cell).

3) Using log–log graph paper show that the following set of values for x and y follows a law of the type $y = ax^n$. From the graph determine the values of the constants a and n.

x	4	16	25	64	144	296
y	6	12	15	24	36	52

4) The following results were obtained in an experiment to find the relationship between the luminosity I of a metal filament lamp and the voltage V:

V	60	80	100	120	140
I	11	20.5	89	186	319

Allowing for the fact than an error was made in one of the readings show that the law between I and V is of the form $I = aV^n$ and find the probable correct value of the reading. Find the value of the constant n.

5) Two quantities t and x are plotted on log–log graph paper, t being plotted vertically and x being plotted horizontally. The result is a straight line and from the graph it is found that:

when $x = 8, \quad t = 6.8$

and when $x = 20, \quad t = 26.9$

Find the law connecting t and x.

6) The intensity of radiation, R, from certain radioactive materials at a particular time t is thought to follow the law $R = kt^n$.

In an experiment to test this the following values were obtained:

R	58	43.5	26.5	14.5	10
t	1.5	2	3	5	7

Show that the assumption was correct and evaluate the constants k and n.

7) The values given in the following table are thought to obey a law of the type $y = ab^{-x}$. Check this statement and find the values of the constants a and b.

x	0.1	0.2	0.4	0.6	1.0	1.5	2.0
y	175	158	60	32	6.4	1.28	0.213

8) The force F on the tight side of a driving belt was measured for different values of the angle of lap θ and the following results were obtained:

F	7.4	11.0	17.5	24.0	36.0
θ rad	$\pi/4$	$\pi/2$	$3\pi/4$	π	$5\pi/4$

Construct a graph to show these values conform approximately to an equation of the form $F = ke^{\mu\theta}$. Hence find the constants μ and k.

9) A capacitor and resistor are connected in series. The current i amperes after time t seconds is thought to be given by the equation $i = I.e^{-t/T}$ where I amperes is the initial charging current and T seconds is the time constant. Using the following values verify the relationship and find the values of the constants I and T:

i amperes	0.0156	0.0121	0.009 45	0.007 36	0.005 73
t seconds	0.05	0.10	0.15	0.20	0.25

SUMMARY

To verify the relationship $z = a.t^n$

rewrite in the form $\log z = n(\log t)+\log a$

and then plot $\log z$ against $\log t$ in order to obtain a straight line graph.

To verify the relationship $z = a.b^t$

rewrite in the form $\log z = (\log b)t+\log a$

and then plot $\log z$ against t in order to obtain a straight line graph.

To verify the relationship $z = a.e^{bt}$

rewrite in the form $\log z = (0.4343b)t+\log a$

and then plot $\log z$ against t in order to obtain a straight line graph.

5. SOLUTION OF EQUATIONS

On reaching the end of this chapter you should be able to:-

1. *Reduce equations to quadratic form in order to find a solution.*
2. *Solve simultaneously a linear and a quadratic equation by an analytical or graphical method.*
3. *Solve simultaneously three first degree equations in three unknowns.*
4. *Use Newton's method to iterate a root of an equation* $f(x) = 0$.

REDUCTION OF EQUATIONS TO QUADRATIC FORM

After making a suitable substitution many equations which are not in quadratic form can be transformed into quadratic form in order to obtain a solution.

EXAMPLE 1

Determine values for z which satisfy $2z^4 - 7z^2 + 6 = 0$.

Let $\quad x = z^2$

$\therefore \quad 2x^2 - 7x + 6 = 0$

$\therefore \quad x = \dfrac{7 \pm \sqrt{49-48}}{4}$

$\therefore \quad x = 1.5 \quad \text{or} \quad x = 2$

$\therefore \quad z^2 = 1.5 \quad \text{or} \quad z^2 = 2$

$\therefore \quad z = \pm\sqrt{1.5} \quad \text{or} \quad z = \pm\sqrt{2}$

$\qquad \approx \pm 1.225 \qquad \approx \pm 1.414$

$\therefore \quad z \in \{-1.414, -1.225, 1.225, 1.414\}$

EXAMPLE 2

Solve the equation $x^{-2/3} - 8x^{-1/3} = 9$.

Let $\quad y = x^{-1/3}$

$\therefore \quad y^2 - 8y - 9 = 0$

$$\therefore \quad y = \frac{8 \pm \sqrt{100}}{2}$$

$$\therefore \quad y = -1 \quad \text{or} \quad y = 9$$

$$\therefore \quad x^{-1/3} = -1 \quad \text{or} \quad x^{-1/3} = 9$$

$$\therefore \quad (x^{-1/3})^{-3} = (-1)^{-3} \quad \text{or} \quad (x^{-1/3})^{-3} = (9)^{-3}$$

$$\therefore \quad x = -\frac{1}{1^3} \quad \text{or} \quad x = \frac{1}{9^3}$$

$$= -1 \qquad\qquad = \frac{1}{729}$$

$$\therefore \quad x \in \left\{-1, \frac{1}{729}\right\}$$

EXAMPLE 3

Calculate values of θ, where $0° \leqslant \theta \leqslant 360°$, such that $3 \cos^2\theta° + 5 \cos \theta° - 2 = 0$.

Let $\quad x = \cos \theta°$

$$3x^2 + 5x - 2 = 0$$

$$x = \frac{-5 \pm \sqrt{25+24}}{6}$$

$$x = -2 \quad \text{or} \quad x = \tfrac{1}{3}$$

$$\cos \theta° = -2 \quad \text{or} \quad \cos \theta° = \tfrac{1}{3}$$

$$\theta \in \varnothing \quad \text{or} \quad \theta \in \{70.5, 289.5\}$$

EXAMPLE 4

Show that if $R = 200$, $c = 160$ and $d = 20$ then the formula $R = \dfrac{c^2}{8(d-h)} - \dfrac{h}{2}$ can be written as a quadratic in h where h is the height of a slip gauge pile. Hence calculate the value of h which satisfies the equation.

Multiplying the equation by the LCM, $8(d-h)$, gives us:

$$8R(d-h) = c^2 - 4h(d-h)$$

or $\quad 4h^2 + (8R - 4d)h + c^2 - 8Rd = 0$

which on substitution becomes:

$$4h^2+1520h-6400 = 0$$

$$h^2+380h-1600 = 0$$

$$h = \frac{-380\pm\sqrt{150\,800}}{2}$$

and taking the positive value

$$h = 4.16$$

Exercise 13

Solve the equations in Questions 1–7 after reducing them to quadratic form by a suitable substitution.

1) $y^4-7y^2+4 = 0$

2) $2\sin^2\theta-3\sin\theta+1 = 0$

3) $x^{1/3}-x^{-1/3} = 1\frac{1}{2}$ (Let $y = x^{1/3}$)

4) $3x^{1/2}+3x^{-1/2} = 10$

5) $144 = 25x^2-x^4$

6) $x^4+400 = 41x^2$

7) $e^{2x} = 5e^x+24$ (Solve for e^x — see Chapter 3.)

8) A long strip of zinc plate, 25 cm wide, is to be made into a gutter by turning strips up vertically along the two sides.
How many centimetres should be turned up to give the rectangular cross-section an area of 70 cm^2?

9) A polygon of M sides always has $\dfrac{M(M-3)}{2}$ diagonals. How many sides has a polygon with 135 diagonals?

10) The plans for a workshop showed the perimeter of a rectangular area to be 324 m. It was decided to increase the length of this area by 6 m and the breadth by 3 m thereby increasing the area by one-ninth of its previous value. Determine the original dimensions.

11) The radius and height of a cylindrical storage tank are both 3 m. The radius must be decreased and the height increased both by the same amount and such that the volume remains the same.
Show that if the radius and height are changed by x m then $x(x^2-3x-9) = 0$ and hence show $x = 0$ is the only solution.

12) Find the radius of a cylinder whose height is 10 cm if the number of square centimetres in its total surface area is equal to the number of cubic centimetres in its volume.

13) Solve the equation $\frac{1}{x^2}+\frac{3}{x}-10 = 0.$

14) A quantity of coal is piled to form a cone 3 m high and on a base of 8 m diameter. The coal is then transferred to a hopper, 8 m long and with a cross-section as shown in Fig. 5.1.
Show that the depth of coal in the hopper is found from $x^2+2x-2\pi = 0$ and hence calculate the depth to the nearest metre.

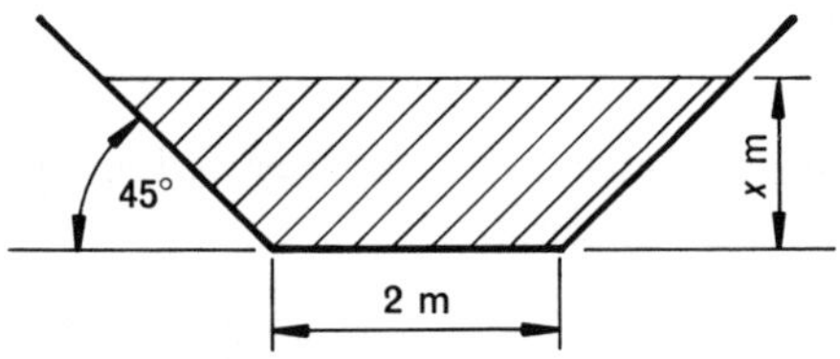

Fig. 5.1

SIMULTANEOUS SOLUTION OF A QUADRATIC AND LINEAR EQUATION

Two methods are available.

1. Substitution Method

This is similar to the method used in Examples 1 and 2 of Chapter 1 of Level II — *Mechanical Engineering Mathematics*.

EXAMPLE 5

Solve simultaneously the equations:

$$y = x^2+3x-4 \qquad [1]$$

and

$$y = 2x+4 \qquad [2]$$

Substituting the value of y given by equation [2], that is $y = 2x+4$, into equation [1] we have:

$$2x+4 = x^2+3x-4$$

from which,

$$x^2+x-8 = 0$$

This equation does not factorise so we will use the formula. Here $a = 1$, $b = 1$ and $c = -8$.

$$\therefore \qquad x = \frac{-1\pm\sqrt{1^2-4(1)(-8)}}{2(1)}$$

$$= \frac{-1\pm\sqrt{33}}{2}$$

$$\therefore \qquad \text{either} \quad x = \frac{-1+5.745}{2} \quad \text{or} \quad x = \frac{-1-5.745}{2}$$

$$\therefore \qquad \text{either} \quad x = 2.372 \quad \text{or} \quad x = -3.372$$

Now for each of these values of x there will be a corresponding value of y. We may find these values of y by substituting the values $x = 2.372$ and $x = -3.372$ into either of the given equations. Therefore using equation [2]:

$$\text{when} \quad x = 2.372, \quad \text{then} \quad y = 2(2.372)+4 = 8.744$$

and

$$\text{when} \quad x = -3.372, \quad \text{then} \quad y = 2(-3.372)+4 = -2.744$$

Hence the required solutions are:

x	2.372	−3.372
y	8.744	−2.744

These solutions may be checked by substituting the values into equation [1]:

$$\text{R.H.S.} = (2.372)^2+3(2.372)-4 = 8.742 = \text{L.H.S. (almost)}$$

and

$$\text{R.H.S.} = (-3.372)^2+3(-3.372)-4 = -2.745 = \text{L.H.S. (almost)}$$

The inaccuracies occur because the values have been rounded off correct to the third place of decimals.

EXAMPLE 6

Solve simultaneously the equations:

$$y^2-2y-4 = x \qquad [1]$$

and

$$x+3y-2 = 0 \qquad [2]$$

Now equation [2] may be rearranged to give $x = -3y+2$ and if we substitute this value of x into equation [1] we have:

$$y^2-2y-4 = -3y+2$$

from which $\quad y^2+y-6 = 0$

and factorising $\quad (y+3)(y-2) = 0$

$\therefore$ either $y+3 = 0$ or $y-2 = 0$

$\therefore$ either $y = -3$ or $y = 2$

Now for each of these values of y there will be a corresponding value of x. We may find these values of x by substituting the values of y into either of the given equations. Therefore using equation [2]:

when $y = -3$, then $x+3(-3)-2 = 0$

$\therefore \quad x = 11$

and when $y = 2$, then $x+3(2)-2 = 0$

$\therefore \quad x = -4$

Hence the required solutions are:

x	11	−4
y	−3	2

These solutions may be checked by substituting the values into equation [1]:

$$\text{L.H.S.} = (-3)^2-2(-3)-4 = 9+6-4 = 11 = \text{R.H.S.}$$

and $\quad \text{L.H.S.} = (2)^2-2(2)-4 = 4-4-4 = -4 = \text{R.H.S.}$

2. Graphical Method

Since the solutions we require have to satisfy both the given equations they will be given by the values of x and y where the graphs of the equations intersect.

The example which follows has already been solved by the substitution method and so we will be able to check the values obtained.

EXAMPLE 7

Solve simultaneously the equations:

$$y = x^2+3x-4$$

and $$y = 2x+4$$

We must first draw up tables of values, and will use the range $x = -4$ to $x = +4$:

x	-4	-2	0	2	4
x^2 $+3x$ -4	16 -12 -4	4 -6 -4	0 0 -4	4 6 -4	16 12 -4
$y = x^2+3x-4$	0	-6	-4	6	24
$2x$ $+4$	-8 4	-4 4	0 4	4 4	8 4
$y = 2x+4$	-4	0	4	8	12

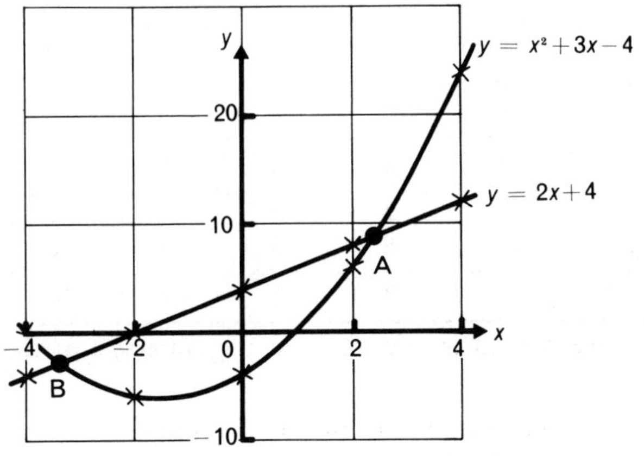

Fig. 5.2

The two graphs are shown plotted on the same axes in Fig. 5.2 and they intersect at the points A and B. Values of the x and y co-ordinates at these points will give the solutions of the given equations. We must be careful not to try to read the values too accurately, as the graphs have been plotted using only five values of x. In this case even values to the first place of decimals cannot be guaranteed.

If more accurate answers are required then we must plot the portions of the graph containing points A and B using more values of x and also much larger scales.

Hence the required solutions are:

x	2.4	−3.4
y	8.7	−2.7

EXAMPLE 8

Solve simultaneously the equations:

$$y^2-2y-6 = x \qquad [1]$$

and

$$x+3y-2 = 0 \qquad [2]$$

In this case it will be easier to draw up tables of values using given values of y and hence find the corresponding x values.

Equation [2] may be rewritten as $x = 2-3y$.

Using values of y from $y = -4$ to $y = +4$ the tables of values will be:

y	−4	−2	0	2	4
y^2	16	4	0	4	16
$-2y$	8	4	0	−4	−8
-6	−6	−6	−6	−6	−6
$x = y^2-2y-6$	18	2	−6	−6	2
2	2	2	2	2	2
$-3y$	12	6	0	−6	−12
$x = 2-3y$	14	8	2	−4	−10

The two graphs are plotted in Fig. 5.3 and they intersect at the points A and B.

As in the previous example the values of the co-ordinates at these points will give the solutions of the given equations.

Again, the accuracy cannot be guaranteed even to the first place of decimals.

The required solutions are:

x	−5.2	12.2
y	2.4	−3.4

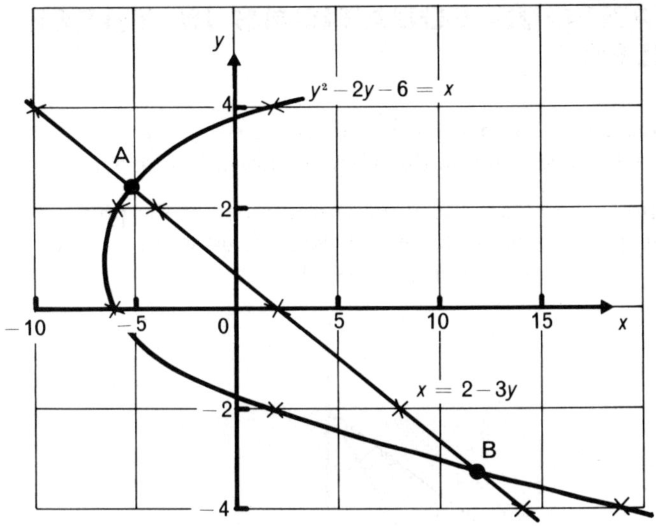

Fig. 5.3

Exercise 14

Solve simultaneously, by substitution:

1) $y = 3x^2 - 3x - 1$
$y = 4x - 3$

2) $y = 8x^2 - 2$
$y = 1 - 5x$

3) $y = 4x^2 - 2x - 1$
$x - y + 1 = 0$

4) $x = y^2 - 0.4y - 1.5$
$x + 0.7y + 0.3 = 0$

5) $y = 7x^2 + 6x - 3$
$2x + y = -1$

6) $y = x^2 + 1.213x + 0.574$
$y = 2.213x + 0.435$

Solve simultaneously, by the graphical method:

7) $x = y^2 - 3y - 2$
$x = 2 - 3y$

8) $y = x^2 - 2x - 2$
$x - y + 2 = 0$

9) $y = x^2 - x + 5$
$y = 2x + 5$

10) $x = 2y^2 - 5y$
$x - 2y - 3 = 0$

11) $y = 5x^2 + x - 3$
$y = 5x - 2$

12) $y = 2x^2 - 2.3x + 1$
$y = 3x - 0.25$

SIMULTANEOUS EQUATIONS IN THREE VARIABLES

The first degree equation in three variables, $2x-3y+4z = 12$ represents a plane. A triangle lying on this plane is shown in Fig. 5.4.

To draw this plane we obtain the three co-ordinates of the points on the axes by letting pairs of variables equal zero. Thus when $y = 0$ and $z = 0$ we have $2x = 12$ leading to (6, 0, 0).

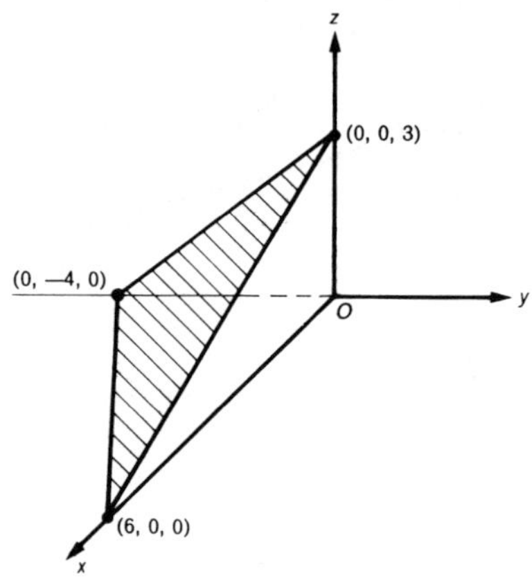

Fig. 5.4

Just as the simultaneous solution of two linear equations can be represented geometrically as the intersection of two lines, so the simultaneous solution of three first degree equations in three variables can be represented geometrically as the intersection of three planes.

We will obtain a unique solution from the three equations if the three planes meet at a point.

To find this point we reduce the three equations in three unknowns to two equations in two unknowns and then solve the remaining pair of equations by the methods described in the previous book.

EXAMPLE 9

Solve the system of equations

$$x-2y+z = 0 \qquad [1]$$

$$4x-3y-2z = 4 \qquad [2]$$

$$3x+4y+2z = 17 \qquad [3]$$

From equation [1] we have $x = 2y-z$.

Substituting into equations [2] and [3] gives

$$4(2y-z)-3y+2z = 4$$

$$3(2y-z)+4y+2z = 17$$

or

$$5y-2z = 4 \qquad [4]$$

$$10y- z = 17 \qquad [5]$$

From equation [4] $\quad y = \frac{2}{5}z+\frac{4}{5}$

Substituting into equation [5] gives

$$10\left(\frac{2}{5}z+\frac{4}{5}\right)-z = 17$$

or

$$3z = 9 \qquad [6]$$

We have thus reduced the initial three equations to the system

$$x-2y+z = 0 \qquad [1]$$

$$5y-2z = 4 \qquad [4]$$

$$3z = 9 \qquad [6]$$

From equation [6] we obtain $\quad z = 3, \quad$ and by back substitution we obtain the solution set $\{(1, 2, 3)\}$; the point where the three planes represented by the initial equations would intersect.

Exercise 15

Solve the following system of equations:

1) $a-5b+c = 0$, $2a+b-c = 0$, $3a+7b+2c = 0$

2) $x+2y-z = -2$, $x-3y+7z = -16$, $x-8y+4z = 3$

3) $x+2y+3z = -1$, $2x-y-z = -4$, $-x+y+3z = 4$

4) $u+v+w = 3$, $3u-v+2w = 4$, $2u+3v+w = 6$

5) $x+y+8z = -3$, $3x+y+32z = -9$, $4x+3y+8z = 2$

6) $2x+y+z = 0$, $3x+y = 1$, $x+z = 1$

7) $3x+2y-z = 6$, $2x-y = 2$, $5x+z = 2$

8) $3x-2y+z = 72$, $x+3y-2z = -48$, $3x+2z = 60$

9) $0.5i+1.1j-0.3k = 6.4$, $1.2i-0.7j+0.8k = 1.7$, $0.9i-1.5j+0.1k = -4.6$

10) The parabola $\quad y = ax^2+bx+c \quad$ passes through the points $(-1, 4)$, $(2, 4)$ and $(1, 2)$. Calculate the values of a, b and c and write down the equation of the parabola.

SOLVING EQUATIONS BY ITERATION

Many equations are such that their roots cannot be found exactly. For example, we cannot find an exact value of x which enables us to solve $x^5-2x+2=0$ or $\tan x - x = 0$.

However, equations of this type arise so frequently in engineering and science that it is important that we are able to calculate the roots of such equations to a stated degree of approximation.

To do this we:

(i) Find an approximation to the root of the required equation. This can be found from a sketch of the curve representing the equation.

(ii) Use this approximation to obtain a better value which we use to find a better value and so on.

NEWTON'S METHOD

Geometrically this method of solving an equation in one variable depends on the fact that the tangent at any point on a curve is a close approximation to the curve for a short distance on either side of the point of contact.

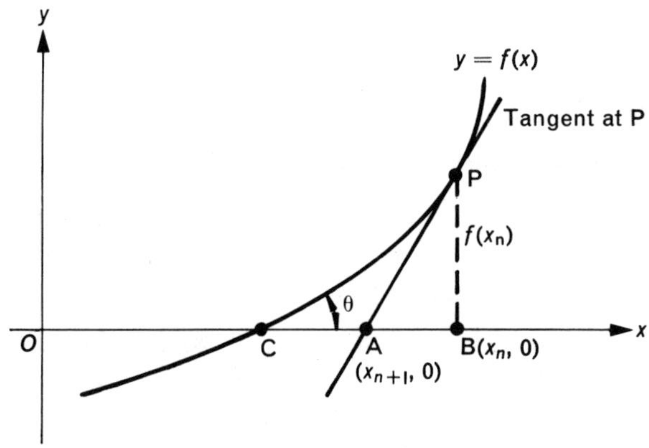

Fig. 5.5

In Fig. 5.5 OC represents the position of the required value of x. $OB = x_n$, a first approximation to OC and $BP = f(x_n)$.

$$\text{AB} = \frac{\text{BP}}{\tan \theta} = \frac{f(x_n)}{f'(x_n)}$$

$$\therefore \qquad OA = OB - AB = x_n - \frac{f(x_n)}{f'(x_n)}$$

which is closer to the required value.

Thus

$$\boxed{x_{n+1} = x_n - \frac{f(x_n)}{f'(x_n)}}$$

EXAMPLE 10

Calculate to three places of decimal the root of $x^3+x-0.5 = 0$ which is near $x = 0$.

Here

$$f(x) = x^3+x-0.5$$

$$f'(x) = 3x^2+1$$

Hence the Newton formula becomes:

$$x_{n+1} = x_n - \frac{x_n^3+x_n-0.5}{3x_n^2+1}$$

From which we obtain, with $x_0 = 0$

$$x_1 = 0 - \frac{0+0-0.5}{0+1} = 0.5$$

$$x_2 = 0.5 - \frac{(0.5)^3+0.5-0.5}{3(0.5)^2+1} = 0.4286$$

$$x_3 = 0.4286 - \frac{(0.4286)^3+0.4286-0.5}{3(0.4286)^2+1} = 0.4239$$

$$x_4 = 0.4239 - \frac{(0.4239)^3+0.4239-0.5}{3(0.4239)^2+1} = 0.4239$$

Hence $x = 0.424$ (3D).

EXAMPLE 11

Calculate to 2D, two places of decimals, the root of $\sin x+x-1 = 0$ near $x = 1$.

Here

$$f(x) = \sin x+x-1$$

$$f'(x) = \cos x+1$$

Hence the Newton formula becomes:

$$x_{n+1} = x_n - \frac{\sin x_n + x_n - 1}{\cos x_n + 1}$$

from which we obtain, with $x_0 = 1$,

$$x_1 = 1 - \frac{0.841+1-1}{0.540+1} = 0.454$$ Remember that $\sin 1 = \sin 57.3^\circ$

$$x_2 = 0.454 - \frac{0.439+0.454-1}{0.899+1} = 0.511$$

$$x_3 = 0.511 - \frac{0.489+0.511-1}{0.872+1} = 0.511$$

Hence $x = 0.51$ (2D).

Exercise 16

Use the Newton process to find the roots of the following equations. In each case a starting value and the degree of approximation is indicated.

1) $3x - \cos x = 0$ $\{x_0 = 0.3;$ to 4D$\}$

2) $x^3 + 2x - 6 = 0$ $\{x_0 = 1.5;$ to 3D$\}$

3) $x^3 - 3x - 5 = 0$ $\{x_0 = 2;$ to 2D$\}$

4) $x - 2 \sin x = 0$ $\{x_0 = 2;$ to 3D$\}$

5) $x^3 - x^2 - x - 3 = 0$ $\{x_0 = 2.5;$ to 3D$\}$

6) Using $x^2 - 6 = 0$ with a starting value of 2.5, determine $\sqrt{6}$ to 4D.

PROJECT

Investigate and/or construct a simple mechanical device which will solve a certain type of equation.

6. SEQUENCES AND SERIES

On reaching the end of this chapter you should be able to:-

1. *Recognise Arithmetic and Geometric sequences and series.*
2. *Add a set number of terms in an Arithmetic or Geometric series.*
3. *Add, when possible, an infinite number of terms in a Geometric series.*
4. *Use sequences and series to solve problems in technology and production.*
5. *Use a sequence of continued fractions to solve practical problems.*

INTRODUCTION

When numbers are arranged according to some rule we say they form a **sequence**. If the sequence is formed by adding a constant to each preceding term then we say it is an **arithmetic** sequence. The first term in an arithmetic sequence is denoted by the letter a, the common difference by d, and if the nth term is denoted by U_n then $U_{n+1} - U_n = U_n - U_{n-1}$. Thus, in general an arithmetic sequence is represented by

$$a,\ a+d,\ a+2d,\ a+3d,\ \ldots,\ a+(n-1)d$$

Note that the coefficient of d in each term is 1 less than the number of the term.

If the sequence is formed by multiplying each preceding term by a constant then we say it is a **geometric** sequence. The first term in a geometric sequence is denoted by the letter a, the common ratio by r and if the nth term is denoted by U_n then $\dfrac{U_{n+1}}{U_n} = \dfrac{U_n}{U_{n-1}}$ Thus, in general, a geometric sequence is represented by

$$a,\ ar,\ ar^2,\ ar^3,\ \ldots,\ ar^{n-1}$$

Note that the index of r in each term is 1 less than the number of the term.

EXAMPLE 1

1500 units of a new product were manufactured on the first day. It is anticipated that the output will increase by 30 per day for the next month. How many units will be produced on the 15th day? $a = 1500$, $d = 30$.

Hence $\qquad$ 15th term $= a+14d$

$= 1500+(14\times 30)$

$= 1920$ units

EXAMPLE 2

A sheet of transluscent plastic, 1 mm thick, absorbs 80% of the light passing through it. How much light is absorbed by a sheet of similar plastic 6 mm thick?

Since each millimetre of plastic transmits 20% of the incident light we require the 6th term of the geometric sequence

$$0.20,\ 0.20^2,\ 0.20^3,\ \ldots$$

i.e. $\qquad$ 6th term $= 0.20^6$

$= 0.000\,064$

Thus 0.0064% of the light is transmitted and 99.9936% is absorbed.

Exercise 17

In each of the following arithmetic sequences calculate the indicated term.

1) Twelfth term in 5, 8, 11, . . .

2) Ninth term in $\frac{3}{4}, \frac{13}{12}, \frac{17}{12}, \ldots$

3) Seventh term in $3\sqrt{2},\ \sqrt{2},\ -\sqrt{2}, \ldots$

Find the missing terms in each of the following arithmetic sequences.

4) 3, —, 7, —, —.

5) 3, —, —, —, 22.

6) —, 8, —, —, –7.

7) 2, —, —, 13, —.

8) Which term of the sequence 2, 9, 16, . . . is 142?

9) Calculate the 7th term of the geometric sequence

$$\frac{3}{64},\ -\frac{3}{16},\ \frac{3}{4},\ -3, \ldots$$

10) The 1st term of a geometric sequence is 3 and the 4th term is 24. Find the common ratio, r.

11) Insert three geometric means between $\frac{27}{8}$ and $\frac{2}{3}$.

12) One third of the air in a tank is removed with each stroke of a vacuum pump. What part of the original amount of air remains in the tank after five strokes?

13) A machine purchased for £2500 depreciates 15% in value every year. Calculate the value of the machine at the end of a four year period.

14) The 4th speed of a shaping machine is 30 strokes per minute and the 6th and highest speed is 120 strokes per minute. Determine the other speeds given that the speeds form a geometric sequence.

SERIES

When the terms of a sequence are added (or subtracted) the sequence is called a **series**. Thus

$$2+4+6+8+10+\dots$$

is an **arithmetic series** with $d = 2$, and

$$2-4+8-16+32-64+\dots$$

is a **geometric series** with $r = -2$.

ARITHMETIC SERIES

We can write down the sum of the first 100 natural numbers as

$$S_{100} = 1+ 2+ 3+\dots+ 99+100$$

or in reverse $$S_{100} = 100+ 99+ 98+\dots+ 2+ 1$$

Adding gives $$2S_{100} = 101+101+101+\dots+101+101$$

$$= 100(101)$$

$$= 10\,100$$

$$\therefore \quad S_{100} = 5050$$

Repeating this process with the general arithmetic series gives

$$S_n = a + (a+d) + (a+2d) + \dots + \{a+(n-2)d\} + \{a+(n-1)d\}$$

$$S_n = \{a+(n-1)d\}+\} a+(n-1)d\} + \{a+(n-3)d\} + \dots + (a+d) + a$$

$$\therefore 2S_n = \{2a+(n-1)d\}+\{2a+(n-1)d\}+\dots+\{2a+(n-1)d\}$$

$$= n\{2a+(n-1)d\}$$

Hence
$$S_n = \frac{n}{2}\{2a+(n-1)d\}$$

If we denote by L the nth or last term in the above series we have

$$L = a+(n-1)d$$

from which
$$n = \frac{L-a}{d}+1$$

EXAMPLE 3

Calculate the sum of the first 12 terms of the arithmetic series $-5-1+3+7+\ldots$

Here $a = -5$, $d = 4$ and $n = 12$.

Hence
$$S_{12} = \frac{12}{2}\{2(-5)+(12-1)4\}$$
$$= 204$$

EXAMPLE 4

Find the sum of the arithmetic series $7+12+17+\ldots+247$.

Here $a = 7$, $d = 5$ and $L = 247$, hence $n = \frac{247-7}{5}+1 = 49$.

$$S_{49} = \frac{49}{2}\{2(7)+(49-1)5\}$$
$$= 6223$$

EXAMPLE 5

A technician has charge of 22 machines, each of which when started goes on working automatically and can produce 4 m of material per hour. He starts the first machine at 9 a.m. and the others at intervals of 5 minutes. What will be the total length of material produced at 1 p.m.?

The first machine will operate for 4 hours or 240 minutes. The second machine for 235 minutes and so on.

Hence total time in minutes $= 240+235+230+\dots$, for 22 terms.

$$= \frac{22}{2}\{2(240)+(22-1)(-5)\}$$

$$= 4125 \text{ min}$$

$$= 68.75 \text{ hours}$$

$\therefore$ Amount of material $= 68.75\times 4 = 275$ m

Exercise 18

1) Find the sum of the first 18 terms of the arithmetic series $4+7+10+\dots$

2) Sum the series $3+8+13+\dots+123$.

3) How many terms of the series $-5-1+3+\dots$ must be added to give a sum of 400?

4) Calculate the sum of all the natural numbers which are exactly divisible by 3 and less than 100.

5) To start a small business a man borrows £54 000. This has to be repaid by annual instalments, but to allow the business to become established only £100 has to be repaid the first year and thereafter each instalment has to be £40 more than the preceding one. In how many years will the debt be cancelled?

6) A technician stacks $(8k+15)$ round fence posts in such a way that there are 8 posts on the top layer and k layers in total. If each layer contains one more post than the layer immediately above, calculate the number of posts.

7) In a strawberry bed, in the shape of a trapezium, a market gardener plants 2457 plants. In each row the number of plants exceeds the number in the preceding row by a fixed amount. How many rows are there if the first row has 77 plants and the last row has 157 plants?

8) A technician working for a certain firm is given a commencing salary of £2400 per annum rising after *two* years service by *annual* increments of £225 to a maximum salary of £6900. How much money will a technician receive during 45 years service with this firm?

9) A set of slip gauges are numbered consecutively 1, 2, 3, ... up to 288. Show that the sum of the slip gauge numbers *before* number 204 equals the sum of slip gauge numbers after number 204.

10) A body falls freely from rest without resistance. It falls 4.9 m during the first second, 14.7 m during the second second, 24.5 m during the third second, and so on. Calculate:

(a) How far it falls during the 7th second.
(b) How far it has fallen altogether in 7 seconds.
(c) How long it will take to fall 490 m.

GEOMETRIC SERIES

Calculate the sum of the first nine terms of the following series:

i.e. $$S_9 = 3+6+12+24+\ldots+384+768$$

We multiply this series by its common ratio, 2, to obtain a second series:

$$2S_9 = \phantom{3+{}} 6+12+24+\ldots+384+768+1536$$

Subtracting: $$S_9 = -3 \qquad\qquad +1536$$
$$= 1533$$

Repeating this procedure using the general geometric series gives:

$$S_n = a+ar+ar^2+ar^3+\ldots+ar^{n-2}+ar^{n-1}$$

$$rS_n = \phantom{a+{}} ar+ar^2+ar^3+\ldots+ar^{n-2}+ar^{n-1}+ar^n$$

$$\therefore \quad S_n - rS_n = a \qquad\qquad -ar^n$$

$$\therefore \quad S_n(1-r) = a(1-r^n)$$

$$\therefore \quad S_n = \frac{a(1-r^n)}{1-r}$$

We see from this result that r cannot equal 1 as we would then have a zero denominator. If r does equal 1 we return to the original series to obtain

$$S_n = a+a+a+\ldots+a$$
$$= na$$

Now $$S_n = \frac{a(1-r^n)}{1-r}$$

can be rewritten as $$S_n = \frac{a}{1-r} - \frac{ar^n}{1-r}$$

and if we make r a proper fraction, that is $-1 < r < 1$, and at the same time we make n very large, that is n tends to infinity, then the second term, $\frac{ar^n}{1-r}$, tends to zero.

Hence for the geometric series we have

$$
\begin{array}{|ll|}
\hline
S_n = \dfrac{a(1-r^n)}{1-r}, & r \neq 1 \\
S_n = na \quad , & r = 1 \\
S_\infty = \dfrac{a}{1-r} \quad , & -1 < r < 1 \\
\hline
\end{array}
$$

EXAMPLE 6

Calculate the sum of the first six terms of the geometric series $5-10-20-\ldots$

Using the above formula we have with $a = 5$, $r = -2$ and $n = 6$:

$$S_6 = \frac{5(1-(-2)^6)}{1-(-2)} = \frac{-315}{3} = -105$$

EXAMPLE 7

Determine the sum to infinity of the series $\frac{1}{2}+\frac{1}{4}+\frac{1}{8}+\ldots$. How many terms of this infinite series must be taken so that the sum may differ from the sum to infinity by less than 0.000 01?

$$S_\infty = \frac{\frac{1}{2}}{1-\frac{1}{2}} = 1 \quad \text{and} \quad S_n = \frac{\frac{1}{2}\{1-(\frac{1}{2})^n\}}{1-\frac{1}{2}} = 1-\frac{1}{2^n}$$

Hence the difference between S_∞ and $S_n = \frac{1}{2^n}$. If this is less than 0.000 01, we have

$$\frac{1}{2^n} < 0.000\,01$$

or

$$-n \log 2 < \log(0.000\,01)$$

$$n > \frac{-\log(0.000\,01)}{\log 2}$$

$$n > 16.6$$

Hence at least 17 terms must be taken.

EXAMPLE 8

Theoretically, when a ball having a coefficient of elasticity of 0.8 is dropped upon a hard surface, it rebounds to a height equal to eight-tenths of the height from which it was dropped. If it is released from a height of 5 m and continues to rebound, how far will it travel before coming to rest.

Let $x =$ distance travelled.

$$x = 5+2(0.8)(5)+2(0.8^2)(5)+\dots$$

$$= 5+10\{0.8+0.8^2+0.8^3+\dots\}$$

The expression in the brackets is a geometric series with $a = 0.8$ and $r = 0.8$.

$$= 5+10\left\{\frac{0.8}{1-0.8}\right\}$$

$$= 45 \text{ m}$$

Exercise 19

Find the last term and the sum of the terms of the following geometric series.

1) $4+8+16+\dots$ to 7 terms.

2) $9+3+1+\dots$ to 11 terms.

3) $2+6+18+\dots$ to 9 terms.

4) $1+\frac{1}{4}+\frac{1}{16}+\dots$ to 10 terms.

5) $1+4+16+\dots$ to 7 terms.

6) $8+2+\frac{1}{2}+\dots$ to 20 terms.

Find the sum to infinity of each of the following series, *if it exists*; if not, say so.

7) $1+\frac{1}{5}+\frac{1}{25}+\dots$

8) $1+\frac{1}{3}+\frac{1}{9}+\dots$

9) $4+1+\frac{1}{4}+\dots$

10) $1+3+9+\dots$

11) $2-\frac{4}{3}+\frac{8}{9}-\frac{16}{27}+\dots$

12) $1-1+1-1+\dots$

13) Find the sum of 10 terms of the geometric series with $a = 1$ and $r = 2$. How many terms must be added to obtain a total of 127?

14) A person contributes £1 to the Red Cross and writes letters to two friends requesting each to contribute £1 and write similar letters to two friends. Assuming that all comply how much would the Red Cross receive after 100 sets of letters were written.

15) The repeating infinite decimal fraction $F = 0.123\,123\ldots$ can be written as $1000F = 123.123\,123\ldots$ Show by subtracting in a similar method to that shown on page 84 that $0.123\,123\ldots = \dfrac{123}{999}$.

16) A square has sides of 20 cm. The midpoints of its sides are joined to form an inscribed square. The midpoints of this second square are joined to form a third square. If the process is continued endlessly, find the sum of the perimeters and the sum of the areas of ali tne squares, including the initial one.

17) A factory introduces a new product at 200 articles in the first week and an increase of 5% each week thereafter until the output reaches 2500 articles. Determine the minimum number of complete weeks required to just exceed this total.

18) A ball is dropped from a height of 32 m. Each time it strikes the ground it rebounds $\frac{3}{8}$ths of the distance from which it has fallen. Theoretically, how far will the ball travel before coming to rest?

19) The yearly output of a mine decreases every year by 13% of its amount during the previous year. Given that the first year's output is worth £2.6 million, find the total output: (a) for the first 10 years, (b) for all time.

CONTINUED FRACTIONS

We can approximate the rational number $\dfrac{60}{53}$ by a series of convergent fractions. From the Level II book we have:

```
53)60(1
   53
   --
   7)53(7
     49
     --
     4)7(1
       4
       -
       3)4(1
         3
         -
         1)3(3
           3
           -
           .
```

$$\frac{60}{53} = 1+\frac{1}{7+}\,\frac{1}{1+}\,\frac{1}{1+}\,\frac{1}{3+}$$

Expressed in a table we obtain the results:

1	7	1	1	3
1	8	9	17	60
1	7	8	15	53

Hence the convergents are $1, \frac{8}{7}, \frac{9}{8}, \frac{17}{15}$ and expressing them as decimal fractions we can calculate their error from the required fraction.

Fraction	Decimal fraction	Error	% error
1	1.000 000	−0.132 076	−13%
$\frac{8}{7}$	1.142 857	+0.010 782	+1.1%
$\frac{9}{8}$	1.125 000	−0.007 075	−0.7%
$\frac{17}{15}$	1.133 333	+0.001 258	+0.1%
$\frac{60}{53}$	1.132 076		

We can also determine the difference between the values of any two consecutive convergents from the reciprocal of the product of their denominators. Thus the difference between $\frac{8}{7}$ and $\frac{9}{8}$ is

$$\frac{1}{7\times 8}=\frac{1}{56}=0.017\,857$$

This simple check determines the nearness of any convergent to the exact value of the given fraction since the deviation from the exact value will be less than this difference from the preceding convergent.

GEAR TRAINS

Screw cutting on a lathe uses the following relationships:

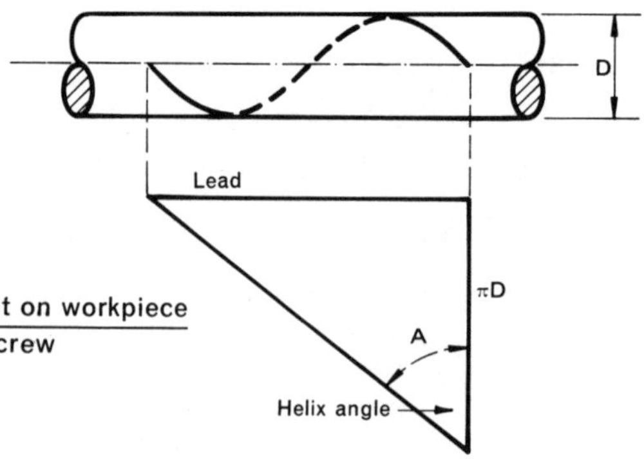

$$\frac{\text{Drivers}}{\text{Driven}}=\frac{\text{Lead of thread to be cut on workpiece}}{\text{Lead of leadscrew}}$$

Having determined the ratio of $\frac{\text{Drivers}}{\text{Driven}}$ the required gear wheels should be obtained from the set available. If the desired ratio is not possible from the set of gears then, by using continued fractions, an approximate ratio should be found.

EXAMPLE 9

A screw of lead 5 mm has to be cut on a lathe having a leadscrew of 6.45 mm. Gears of 20T to 120T, in steps of 5T, are available.

Evaluate an approximate gear train and the resulting error, in millimetres, over 20 leads.

$$\frac{\text{Lead to be cut}}{\text{Lead on leadscrew}} = \frac{5\text{ mm}}{6.45\text{ mm}} = \frac{100}{129}$$

	1	3	2	4	3
0	1	3	7	31	100
1	1	4	9	40	129

```
100)129(1
    100
     29)100(3
         87
         13)29(2
            26
             3)13(4
               12
                1)3(3
                  3
                  -
                  .
```

The closest convergent to $\frac{100}{129}$ is $\frac{31}{40}$ but this cannot be accommodated by the gears available.

The next approximation, $\frac{7}{9}$, can be obtained using $\frac{7\times 5}{9\times 5} = \frac{35\text{T}}{45\text{T}}$.

20 leads should be $\quad 20\times 5\text{ mm} = 100\text{ mm}$

but using the approximate gear set we have

$$20\text{ leads} = 20\times\frac{7}{9}\times 6.45$$

$$= 100.\dot{3}\text{ mm}$$

Hence using $\frac{35\text{T}}{45\text{T}}$, the error over 20 leads is $\frac{1}{3}$ mm.

EXAMPLE 10

Two spur gears have to provide a ratio close to 3.21 and have to mesh at a centre distance of 120 mm. Use continued fractions to determine a convenient metric module and the numbers of teeth on each spur gear.

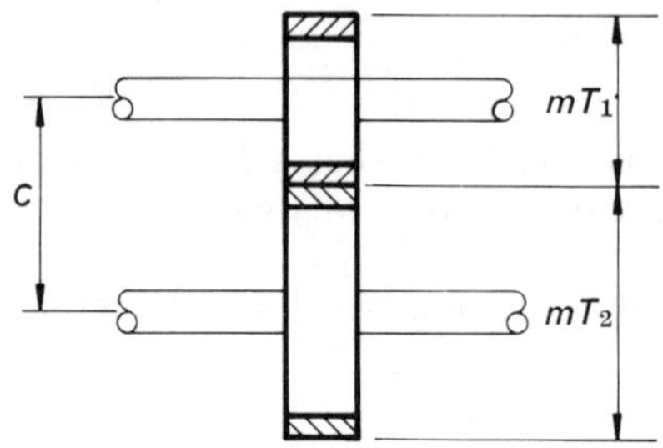

The centre distance, C, of spur gears having T_1 and T_2 teeth of metric module m, is given by

$$C = \tfrac{1}{2}m(T_1+T_2)$$

Obviously T_1 and T_2 must be whole numbers and to avoid using non-standard cutters we also require m to be a whole number.

Calculating the convergents for 3.21, or $\frac{321}{100}$ we obtain:

3	4	1	3	5
3	13	16	61	321
1	4	5	19	100

Adding numerator and denominator of all the convergents we have

$$T_1+T_2 = 4,\ 17,\ 21,\ 80,\ 421$$

But $m = \dfrac{2C}{T_1+T_2} = \dfrac{240}{T_1+T_2}$ and to ensure m is a whole number we take

$T_1+T_2 = 80$. That is, we use the convergent $\frac{61}{19}$.

Hence module $= \dfrac{240}{80} = 3$ mm with gears having 61T and 19T.

ANGULAR SPACING WITH A DIVIDED HEAD

Angular spacing between points on a workpiece can be accurately set using a divided head. This has a standard ratio of 40 to 1; that is, 40 revolutions of the index arm will cause the workpiece to rotate through 360°, or one revolution of the arm will rotate the workpiece through $\frac{360°}{40}$ or 9°.

$$\text{Turns of index arm} = \frac{\text{Angular spacing}}{9°}$$

EXAMPLE 11

Calculate, using continued fractions, the nearest indexing ratio which will rotate a workpiece through 6.83°. The index plate has hole circles from 30 holes to 60 holes inclusive, at unit intervals.

$$\text{Number of turns} = \frac{\text{Angular spacing}}{9°} = \frac{6.83°}{9°} = \frac{683}{900}$$

Calculating the convergents, we obtain:

0	1	3	6	1	3	1	1	3
0	1	3	19	22	85	107	192	683
1	1	4	25	29	112	141	253	900

From the 4th convergent, $\frac{22}{29}$, we can use 44 holes on the 58 hole circle to obtain $\frac{44}{58} \times 9° \approx 6.827°$ with an error of 0.0024°.

Exercise 20

1) Calculate a sequence of convergents to the fractions

(a) $\frac{29}{67}$ (b) $\frac{5855}{8477}$ (c) $\frac{147}{62}$

2) A screw of lead 5.5 mm has to be cut on a lathe having a leadscrew of 6.51 mm. Gears of 20T to 150T, in steps of 5T are available. Show, using continued fractions, that the 4th convergent can be obtained using the compound gear train $\frac{35T \times 70T}{145T \times 20T}$ and determine the resulting error, in millimetres, over a length of 10 leads.

3) Two spur gears have to mesh at a centre distance of 100 mm and provide a gear ratio close to 1.27. Show, using continued fractions, that a possible approximation is obtained using $\frac{70T}{55T}$ with a module of 4 mm.

4) A lathe has a leadscrew of 10 mm lead and a set of gear wheels ranging from 20T to 120T in steps of 5T. Determine suitable gear wheels for cutting a screw thread having a lead of 7.40 ± 0.02 mm.

5) Determine, by the method of continued fractions, the closest approximation that can be obtained for an angular indexing of 4.31°. The hole circles on the indexing plate are:

24, 25, 28, 30, 34, 37, 38, 39, 41, 42, 43, 45, one side, and
46, 47, 49, 51, 53, 54, 57, 58, 59, 62, 66, 71, other side.

6) A dividing head has a 40 to 1 worm gear ratio and the index plate has every hole circle from 30 holes to 60 holes per circle inclusive. Determine:

(a) The nearest approximate hole spacing and hole circle to achieve an angle of 8.27° (i.e. 8° 16′ 12″).

(b) The amount of error, to the nearest second, that results.

PROJECTS

1) Select a piece of equipment and investigate when it should be replaced. Establish various sequences related to it, for example capital cost, maintenance etc., and calculate the best replacement time. (Source: Any introductory book on Operations Research.)

2) Investigate Time Series. Use the sales figures for the last few years of a particular company and using moving averages calculate the trend line.

3) Investigate the use of continued fractions applied to the cutting of cams on a vertical milling machine.

4) Investigate the use of continued fractions applied to cutting leads on a hobbing machine with a differential attachment.

7. TRIGONOMETRY

er reaching the end of this chapter you should be able to:-

Solve trigonometrical equations.
Determine the modified radius of a radial flat tool.
Determine the modified shape of a single point flat tool.
4. *Determine the modified shape of a circular form tool.*
5. *Solve problems using vector algebra.*

EQUATIONS

The solution of a trigonometrical equation eventually involves the determination of a single ratio, say $\sin \theta$, from which we can calculate the possible values of the angle θ.

EQUATIONS LIKE $\sin \theta =$ constant

EXAMPLES 1

(a) Determine the values of θ which satisfy $\sin \theta° = 0.400$.

Using a calculator or mathematical tables we obtain

$$\sin \theta° = 0.400$$

$$\Rightarrow \qquad \theta° = 23.58°$$

However, from consideration of Fig. 7.1(a) or (b) we see that, within the limits $0 < \theta < 360$, and from the *symmetry* of the curve, that $\theta°$ also equals 156.42°.

i.e. $\theta \in \{23.58, 156.42\}$.

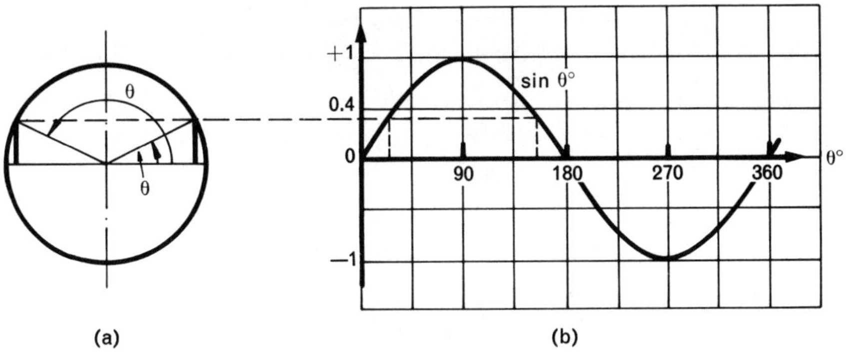

Fig. 7.1

(b) Determine the values of θ which satisfy $\tan\theta° = 0.800$ where $0 < \theta < 360$.

Using a calculating aid we have:

$$\tan\theta° = 0.800$$

$$\Rightarrow \qquad \theta° = 38.66°$$

However, from consideration of Fig. 7.2 we see that for $0 < \theta < 360$, the *symmetry* of the function $\tan\theta$ shows that $\theta°$ can also have the value of 218.66°.

Hence $\qquad \theta \in \{38.66, 218.66\}$

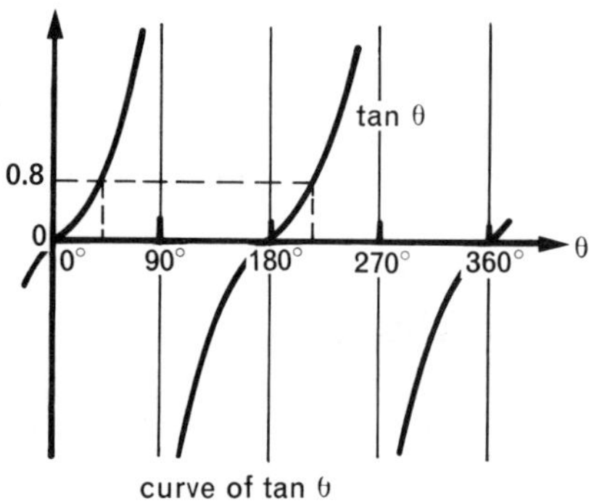

Fig. 7.2

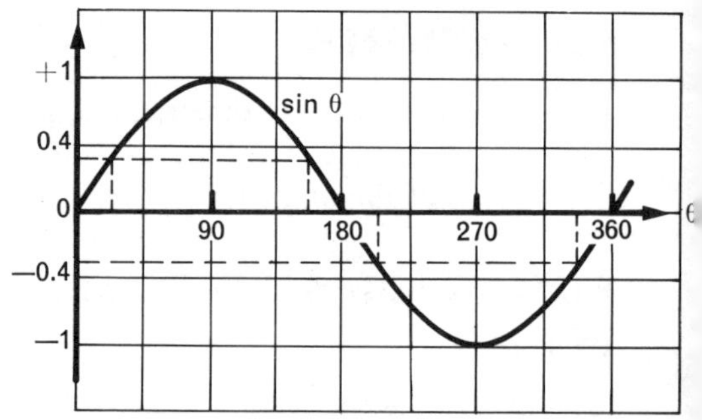

Fig. 7.3

EQUATIONS LIKE $\sin^2\theta$ = constant

EXAMPLES 2

(a) Calculate values of θ, where $0 < \theta < 360$, which satisfy $\sin^2\theta° = 0.160$.

From a calculating aid:

$$\sin^2\theta° = 0.160$$

$$\Rightarrow \qquad \sin\theta° = \pm 0.400$$

Fig. 7.3 shows that there are four possible values of θ, where $0 < \theta < 360$, which satisfy the equation. From Example 1(a):

$$\sin\theta° = 0.400$$

$$\Rightarrow \qquad \theta \in \{23.58, 156.42\}$$

Using the *symmetry* of Fig. 7.3 we can calculate:

$$\sin\theta° = -0.400$$

$$\Rightarrow \qquad \theta \in \{203.58, 336.42\}$$

Hence $\qquad \theta \in \{23.58, 156.42, 203.58, 336.42\}$

(b) Solve the equation $2\cos^2\theta° = 0.720$ where $0 < \theta < 360$.

$$2\cos^2\theta° = 0.720$$

$$\Rightarrow \quad \cos^2\theta° = 0.360$$

$$\Rightarrow \quad \cos\theta° = \pm 0.600$$

$$\Rightarrow \quad \theta° = 53.13°$$

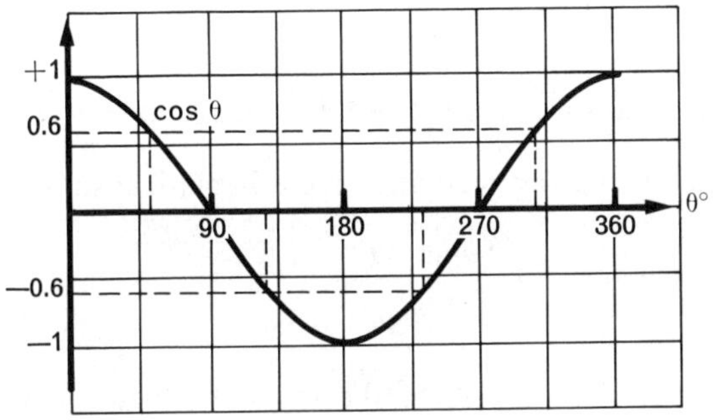

Fig. 7.4

and from the *symmetry* of Fig. 7.4 we have:

$$\theta \in \{53.13, 126.87, 233.13, 306.87\}$$

Exercise 21

Solve the following equations for all values of the variable between 0° and 360°.

1) $\cos\theta° = 0.6$

2) $\sin\theta° = 0.25$

3) $\tan\theta° = -0.7$

4) $\sin\theta° = -\dfrac{1}{\sqrt{2}}$

5) $\sqrt{3}\tan\theta° = -1$

6) $2\sin\theta° + 1 = 0$

7) $\sin^2\theta° - 0.09 = 0$

8) $\cos^2\theta° = 0.04$

9) $\tan^2\theta° = 1.44$

EQUATIONS LIKE $\sin 3\theta = $ constant

EXAMPLES 3

(a) Find all the values of x which satisfy $\sin 2x° = 0.500$, where $0 < x < 360$.

$$\sin 2x° = 0.500$$

$$\Rightarrow \quad 2x° = 30° \quad \text{or} \quad 150°$$

Since the sine curve has a period of 360°, we can rewrite this as:

$$2x^\circ = 30^\circ + n.360^\circ \quad \text{or} \quad 150^\circ + n.360^\circ \quad \text{where} \quad n \in Z$$

i.e. $2x \in \{30+n.360\} \cup \{150+n.360\}$

$\Rightarrow \quad x \in \{15+n.180\} \cup \{75+n.180\}$

$\Rightarrow \quad x \in \{15, 195\} \cup \{75, 255\}$

$\Rightarrow \quad x \in \{15, 75, 195, 255\}$

(b) Solve the equation $\tan 3\theta^\circ = 1.6$ where $0 < \theta < 360$.

Using a calculating machine or mathematical tables we have

$$\tan 3\theta^\circ = 1.6$$

$$\Rightarrow \quad 3\theta^\circ = 58^\circ$$

From Fig. 7.2, the graph of the tangent function, we see that the period of repetition is every 180°. Hence

$$3\theta^\circ = 58^\circ + n.180^\circ \quad \text{where} \quad n \in Z.$$

$$\Rightarrow \quad \theta = 19.3 + n.60$$

$$\Rightarrow \quad \theta \in \{19.3, 79.3, 139.3, 199.3, 259.3, 319.3\}$$

EQUATIONS LIKE $\sin(\theta+30)^\circ = $ constant

EXAMPLE 4

Solve the equation $\sin(\theta+30)^\circ = 0.951$, $0 < \theta < 360$.

$$\sin(\theta+30)^\circ = 0.951$$

$$\Rightarrow \quad (\theta+30)^\circ = 72^\circ \quad \text{or} \quad 108^\circ$$

$$\Rightarrow \quad \theta = 72-30 \quad \text{or} \quad 108-30$$

$$= 42 \quad \text{or} \quad 78$$

$$\Rightarrow \quad \theta \in \{42, 78\}$$

In certain cases values beyond 360° must be considered initially since after subtraction the value of the angle could then be in the appropriate range. The initial value obtained could also become negative after subtraction and would then be omitted since it would be outwith $0 < \theta < 360$.

Exercise 22

Solve the following equations for all values of the variable between 0° and 360°.

1) $\sin 4\theta^\circ = 0.5$

2) $\cos 3x^\circ = -0.5$

3) $\tan 3\theta^\circ - 1 = 0$

4) $\sin(30+\theta)^\circ = 0.891$

5) $\cos(75+\theta)^\circ = -0.682$

6) $\cos(x-60)^\circ = 0.969$

7) $\tan(x-45)^\circ - 0.601 = 0$

QUADRATIC EQUATIONS

Many trigonometric equations are in quadratic form or can be changed into quadratic form using the Pythagorean relation $\sin^2\theta + \cos^2\theta = 1$.

EXAMPLE 5

Solve the equation $2\cos^2 x^\circ + \sin x^\circ = 1$ where $0 < x < 360$.

$$2\cos^2 x^\circ + \sin x^\circ = 1$$

To make this equation a quadratic in $\sin x$, we replace $\cos^2 x$ with $1-\sin^2 x$.

Hence $$2(1-\sin^2 x) + \sin x^\circ = 1$$

or $$2\sin^2 x^\circ - \sin x^\circ - 1 = 0$$

Using the quadratic formula with $a = 2$, $b = -1$ and $c = -1$ we have

$$\sin x^\circ = \frac{1 \pm \sqrt{(1+8)}}{4} = 1 \quad \text{or} \quad -\tfrac{1}{2}$$

We now have two simple equations to solve:

$$\sin x^\circ = 1 \quad \text{or} \quad \sin x^\circ = -\tfrac{1}{2}$$

$$\Rightarrow \quad x^\circ = 90^\circ \quad \text{or} \quad x^\circ = 210^\circ, 330^\circ$$

$$\Rightarrow \quad x \in \{90, 210, 330\}$$

EQUATIONS LIKE $a\sin\theta + b\cos\theta =$ constant

EXAMPLE 6

Solve the equation $7\cos\theta^\circ + 4\sin\theta^\circ = 6$, $0 < \theta < 360$.

Let $$7\cos\theta^\circ + 4\sin\theta^\circ = k\cos(\theta - a)^\circ$$
$$= k\cos\theta^\circ\cos a^\circ + k\sin\theta^\circ\sin a^\circ$$

Then $\left.\begin{aligned} k\cos a^\circ &= 7 \\ k\sin a^\circ &= 4 \end{aligned}\right\}$ hence $k = \sqrt{(7^2+4^2)} = \sqrt{65}$

and $$\tan a^\circ = \frac{4}{7} \text{ in the first quadrant.}$$

$\Rightarrow$ $$a^\circ = 29.8^\circ$$

Hence $$7\cos\theta^\circ + 4\sin\theta^\circ = 6$$

$\Rightarrow$ $$\sqrt{65}\cos(\theta-29.8)^\circ = 6$$

$\Rightarrow$ $$\cos(\theta-29.8)^\circ = \frac{6}{\sqrt{65}} = 0.744$$

$\Rightarrow$ $$(\theta-29.8)^\circ = 41.9^\circ \quad \text{or} \quad 318.1^\circ$$

$\Rightarrow$ $$\theta \in \{71.7,\ 347.9\}$$

Exercise 23

Determine the values of the variable between 0° and 360° which satisfy the following equations.

1) $3\cos^2 x^\circ + 2\cos x^\circ - 1 = 0$
2) $\tan^2\theta^\circ - 3\tan\theta^\circ - 4 = 0$
3) $2 - 3\cos\theta^\circ - 5\cos^2\theta^\circ = 0$
4) $6\cos^2 x^\circ + 7\sin x^\circ - 8 = 0$
5) $6\sin^2 x^\circ - 7\cos x^\circ - 8 = 0$
6) $10\cos^2\theta^\circ - \sin\theta^\circ - 7 = 0$
7) $3\cos\theta^\circ + 4\sin\theta^\circ = 5$
8) $9\cos\theta^\circ + 40\sin\theta^\circ = 0$
9) $2\cos x^\circ - \sin x^\circ = \sqrt{5}$
10) $\sin\theta^\circ + \cos\theta^\circ = -1$

TOOL FORMS

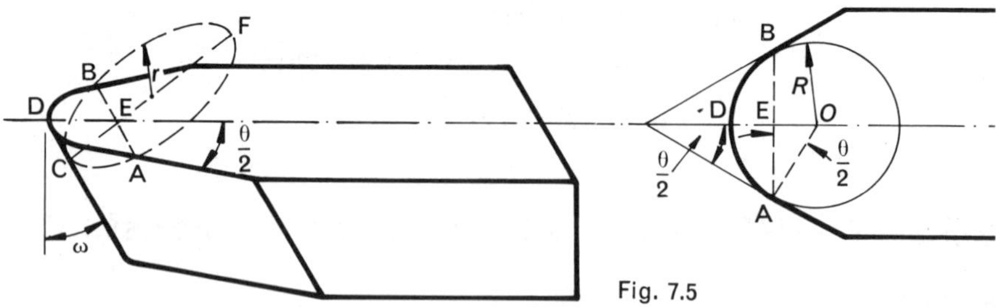

Fig. 7.5

The radial flat tool, Fig. 7.5, has a cutting edge which is the portion of a circle of radius R, an angle of θ° between the tangents drawn to the extremities of the cutting arcs, and a tool clearance angle, ω.

To obtain the required radius R, which has to be imparted to the work-piece, we have to grind a modified radius r perpendicular to the clearance angle, ω. This latter section is really a portion of an ellipse but may be considered as a circle, with little error, while DE is less than half the radius R.

$$\text{AE} = R\cos\tfrac{1}{2}\theta$$

$$\Rightarrow \qquad \text{AB} = 2R\cos\tfrac{1}{2}\theta$$

$$\text{DE} = R - R\sin\tfrac{1}{2}\theta$$

$$\text{EC} = \text{DE}\cos\omega$$

$$\text{EF}\times\text{EC} = \tfrac{1}{2}\text{AB}\times\tfrac{1}{2}\text{AB}$$

$$\Rightarrow \qquad \text{EF} = \frac{\text{AB}^2}{4\text{EC}}$$

$$\Rightarrow \qquad r = \tfrac{1}{2}\left(\frac{\text{AB}^2}{4\text{EC}} + \text{EC}\right) \qquad [1]$$

Hence given the required cutting radius R, the tangential angle θ and the clearance angle ω, we can calculate the modified radius r at which the tool must be ground.

In order to make this tool, it is also necessary to determine the angle between the two angular faces of the tool. The necessary formula is determined from the single point tangential tool form of Fig. 7.6.

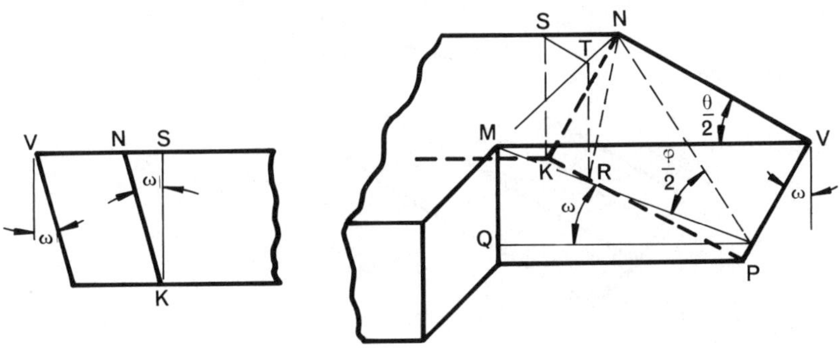

Fig. 7.6

Fig. 7.6 shows the front elevation and a 'cut away' pictorial view of a single point tangential tool form. To achieve a cutting angle of $\theta°$, with a front clearance angle of $\omega°$, we have to grind at an angle of $\phi°$.

Angle MVN $= \tfrac{1}{2}\theta°$, angle MPQ $= \omega°$ and angle MPN $= \tfrac{1}{2}\phi°$.

MN $=$ MV $\tan\tfrac{1}{2}\theta°$ and MP $=$ MV $\cos\omega°$.

Hence
$$\tan \text{MPN} = \frac{\text{MN}}{\text{MP}}$$
$$\tan \tfrac{1}{2}\phi^\circ = \frac{\tan \frac{1}{2}\theta^\circ}{\cos \omega^\circ} \qquad [2]$$

To determine the side clearance angle TRN, β, we have
$$\tan \omega^\circ = \frac{\text{NS}}{\text{KS}} = \frac{\text{NS}}{\text{TR}}$$
$$\text{NS} = \text{TR} \tan \omega^\circ$$
Also
$$\text{TN} = \text{NS} \sin \tfrac{1}{2}\theta^\circ$$
and
$$\tan \text{TRN} = \frac{\text{TN}}{\text{TR}}$$
$$\tan \beta^\circ = \frac{\text{NS} \sin \frac{1}{2}\theta^\circ}{\text{TR}}$$
$$\tan \beta^\circ = \tan \omega^\circ \sin \tfrac{1}{2}\theta^\circ \qquad [3]$$

Equations [1], [2] and [3] give us sufficient information to grind tangential radial and single point form tools.

EXAMPLE 7

Determine the radius r of the approximate circle perpendicular to the clearance line of the radial flat tool shown in Fig. 7.7.

$\frac{1}{2}\theta^\circ = 44^\circ$ and $\omega^\circ = 15^\circ$.

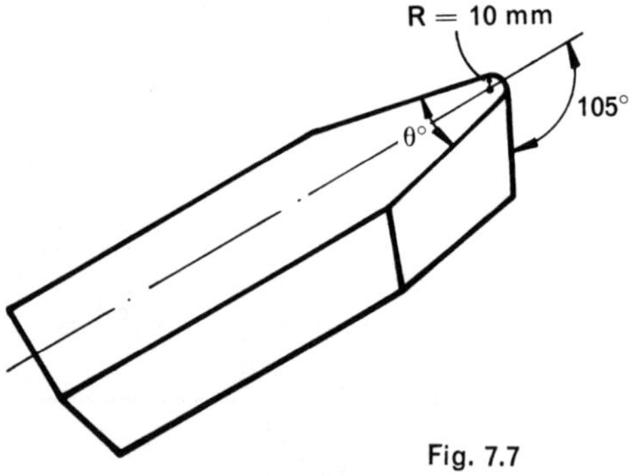

Fig. 7.7

From equation [1] we have:
$$\text{AB} = 2R \cos \tfrac{1}{2}\theta^\circ = 2 \times 10 \times 0.719 = 14.39$$
$$\text{DE} = R - R \sin \tfrac{1}{2}\theta^\circ = 3.05$$
$$\text{EC} = \text{DE} \cos \omega^\circ = 3.05 \times 0.966 = 2.946$$
$$r = \tfrac{1}{2}\left(\frac{\text{AB}^2}{4\text{EC}} + \text{EC}\right) = \tfrac{1}{2}(17.57 + 2.946)$$
$$= 10.23 \text{ mm}$$

EXAMPLE 8

A single point tangential tool form has a 12° front clearance angle and is to be used to cut a 60° vee-groove in a workpiece. Calculate:

(a) the angle to which the tool is ground perpendicular to the clearance line.
(b) the side clearance angle.

(a)
$$\tan \tfrac{1}{2}\phi^\circ = \tan \tfrac{1}{2}\theta^\circ \times \frac{1}{\cos \omega^\circ} = \tan 30^\circ \times \frac{1}{\cos 12^\circ}$$
$$= 0.5774 \times 1.0223 = 0.5903$$

Hence $\tfrac{1}{2}\phi^\circ = 30.5533^\circ$

$\therefore$ $\phi^\circ = 61.12^\circ$ or $61^\circ\ 06'\ 24''$

(b)
$$\tan \beta^\circ = \tan \omega^\circ \times \sin \tfrac{1}{2}\theta^\circ = \tan 12^\circ \times \sin 30^\circ$$
$$= 0.2126 \times 0.5000 = 0.1063$$

$\therefore$ $\beta^\circ = 6.07^\circ$ or $6^\circ\ 03'\ 59''$

CIRCULAR FORM TOOLS

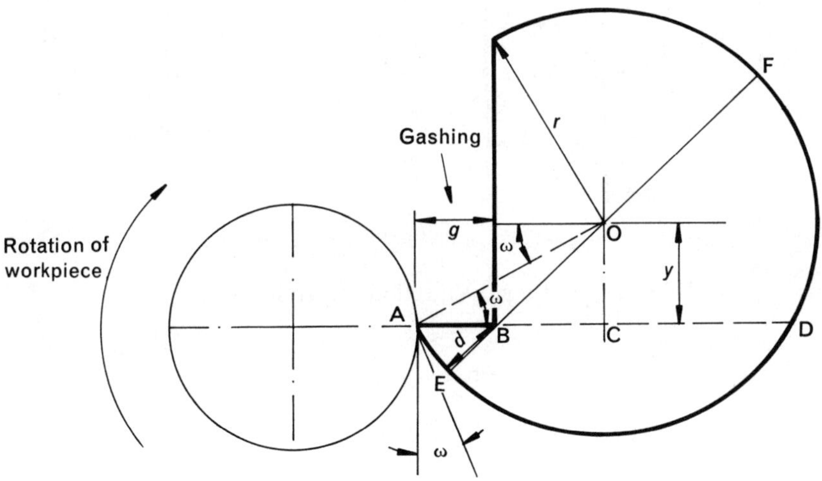

Fig. 7.8

Fig. 7.8 shows a circular form tool. The cutting clearance is obtained by making the cutting edge, AB, a distance y below the centre. The clearance angle, ω°, is the angle CAO. Hence

$$\sin \omega^\circ = \frac{y}{\text{AO}} = \frac{y}{\text{Tool radius } r}$$

$$\therefore \qquad y = r \sin \omega^\circ \qquad [1]$$

A radial depth EB $= d$ will turn a depth AB $= g$ on the workpiece, the width being unaffected. Hence

$$\text{EB.BF} = \text{AB.BD}$$

$$\Rightarrow \qquad d(2r-d) = g(2r \cos \omega^\circ - g)$$

$$\Rightarrow \qquad d^2 - 2rd + g(2r \cos \omega^\circ - g) = 0 \qquad [2]$$

This is a quadratic equation in which all quantities except d are known.

EXAMPLE 9

A circular form tool, 75 mm diameter, is to be made to produce the form shown in Fig. 7.9(a). The diameter of the tool is 75 mm and it is gashed so that its front cutting edge is 7 mm below the tool centre. Determine the shape of a radial section.

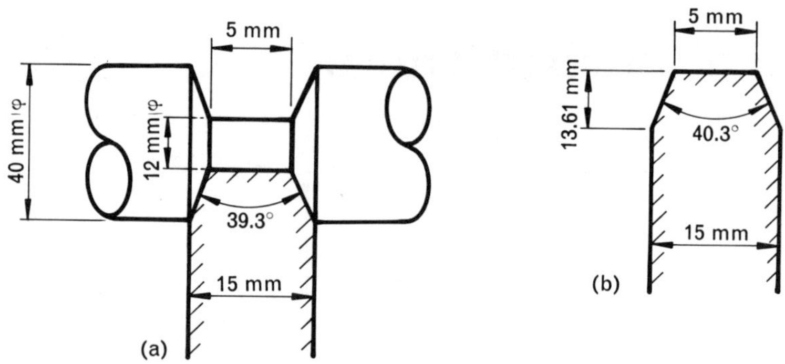

Fig. 7.9

$r = \dfrac{75}{2} = 37.5$ mm and using equation [1] we have

$$\sin \omega^\circ = \frac{7}{37.5}$$

$$\therefore \qquad \omega^\circ = 10.76^\circ$$

$$\therefore \qquad \cos \omega^\circ = 0.9824$$

$g = \dfrac{40-12}{2} = 14$ mm and using equation [2] we have

$$d^2 - 75d + 14(59.68) = 0$$

from which the practical value of d is 13.61 mm.

The angle on the tool to give an angle of 39.3° on the workpiece is found from

$$\tan \tfrac{1}{2}\theta = \frac{5}{13.61}$$

from which $\theta = 40.3°$

The required profile of the tool is shown in Fig. 7.9(b).

Exercise 24

1) Determine the radius r of the approximate circle perpendicular to the clearance line of the radial flat tool with a 15° front clearance angle, a plan angle of 80° between the tangents and a top radius of 10 mm.

2) A single point tangential form tool with a 15° front clearance angle and zero rake is to be used to cut a 55° vee-groove in a workpiece. Calculate:

(a) the angle to which the tool is ground perpendicular to the clearance line,

(b) the side clearance angle.

3) A circular form tool with zero top rake has a cutting diameter of 80 mm. The centre line of the workpiece is offset from the centre line of the tool by 5 mm. Calculate the clearance angle at the leading edge.

4) A single point tangential form tool with a 10° front clearance angle and zero rake angle, is to be used to cut a 58° vee-groove in a cylindrical workpiece. Determine:

(a) the included angle to which the tool must be ground,

(b) the side clearance angle.

5) A circular form tool, 60 mm diameter, is gashed to give 10° clearance, and is to turn the form indicated in Fig. 7.10. Calculate:

(a) the distance the cutting edge is off-centre,

(b) the radial distance XY on the tool.

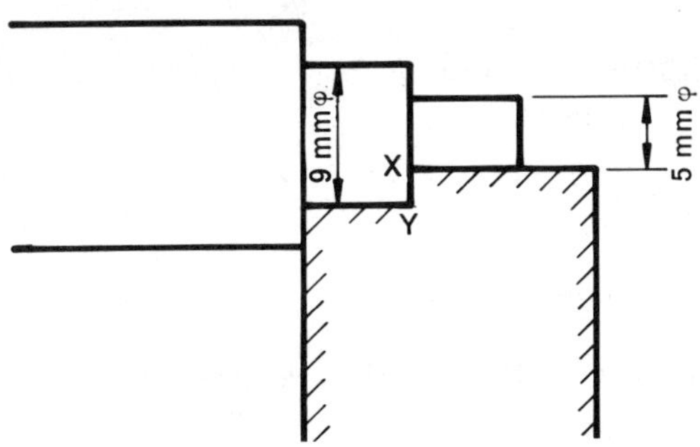

Fig. 7.10

6) A circular form tool of width 15 mm and 100 mm outside diameter is to be used to cut a symmetrical vee-groove of 90° included angle on a cylindrical workpiece. The clearance angle is 10°. Determine:

(a) the offset distance between the centre of the form tool and the centre of the workpiece,
(b) the included angle of a radial section through the tool.

VECTOR ALGEBRA

Quantities which are defined by both magnitude and direction are called **vectors**. They occur frequently in all branches of engineering, and so an understanding of vector algebra is important.

The following points are reminders.

(i) A vector in three dimensions can be represented by one member from a set of directed line segments such as **AB**. It can be defined in component form by the number triple $\begin{pmatrix} x \\ y \\ z \end{pmatrix}$.

(ii) Every vector can be expressed uniquely in terms of three non-coplaner vectors which form the basis for the whole set of vectors. The basis used with rectangular coordinates is **i**, **j** and **k**, vectors of unit length in the directions of the axes Ox, Oy and Oz respectively.

Thus if $\mathbf{a} = \begin{pmatrix} x \\ y \\ z \end{pmatrix}$ then $\mathbf{a} = x\mathbf{i}+y\mathbf{j}+z\mathbf{k}$.

(iii) The position vector of a point P dividing AB in the ratio $m:n$ is given by

$$\mathbf{p} = \frac{m\mathbf{a}+n\mathbf{b}}{m+n}$$

If P divides the line joining $A(x_1, y_1, z_1)$ and $B(x_2, y_2, z_2)$ in the ratio $m:n$, then

$$x_p = \frac{mx_2+nx_1}{m+n}, \quad y_p = \frac{my_2-ny_1}{m+n}, \quad z_p = \frac{mz_2+nz_1}{m+n}$$

the *section formulae*.

(iv) $|\mathbf{AB}| = |\mathbf{b}-\mathbf{a}| = \sqrt{\{(x_2-x_1)^2+(y_2-y_1)^2+(z_2-z_1)^2\}}$; the *distance formula*.

(v) The scalar (dot) product of two vectors **a** and **b** is defined by

$$\mathbf{a.b} = |\mathbf{a}||\mathbf{b}| \cos\theta \quad \text{or} \quad \mathbf{a.b} = a_1b_1 + a_2b_2 + a_3b_3$$

and $$\cos\theta = \frac{\mathbf{a.b}}{|\mathbf{a}||\mathbf{b}|} = \frac{a_1b_1+a_2b_2+a_3b_3}{\sqrt{(a_1^2+a_2^2+a_3^2)(b_1^2+b_2^2+b_3^2)}}$$

also $\mathbf{a.b} = 0 \Rightarrow 0^\circ = 90^\circ$, if **a** and **b** are non-zero vectors.

PROJECTION ON A LINE

The projection of line segment AB on XY is line segment CD.

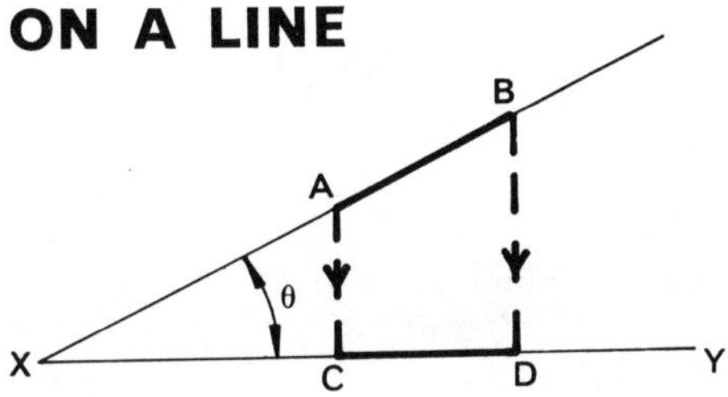

Fig. 7.11

If $\mathbf{AB} = \mathbf{a}$ and the unit vector along **XY** is denoted by **u** then the magnitude of **CD** is $|\mathbf{u}||\mathbf{a}| \cos\theta$. That is $|\mathbf{a.u}|$.

Hence the *resolved* part of **a** in the direction of **u** is **a.u**.

EXAMPLE 10

A ship is travelling at 15 knots on a bearing of 030°. Represent this vector quantity by **v** and find the resolved part due north. Calculate the time taken for the ship to change its latitude by one degree. (1° of latitude = 60 nautical miles.)

$$\begin{aligned}\text{Resolved part due north} &= \mathbf{v.n}\\ &= |\mathbf{v}||\mathbf{n}| \cos 30^\circ\\ &= 15\times1\times0.866\\ &\approx 13 \text{ knots}\end{aligned}$$

Hence it would take $\frac{60}{13}$ hours to change latitude by 1°. That is 4 hr 37 min.

EXAMPLE 11

The points A(3, 4, 6), B(4, 2, 2) and C(6, f, g) are collinear. Determine the values of f and g and the ratio in which B divides AC.

$$\mathbf{a} = \begin{pmatrix}3\\4\\6\end{pmatrix}, \quad \mathbf{b} = \begin{pmatrix}4\\2\\2\end{pmatrix} \quad \text{and} \quad \mathbf{c} = \begin{pmatrix}6\\f\\g\end{pmatrix}$$

For the points to be collinear $\mathbf{b}-\mathbf{a} = k(\mathbf{c}-\mathbf{b})$, where k is a constant.

Hence $$\begin{pmatrix}1\\-2\\-4\end{pmatrix} = k\begin{pmatrix}2\\f-2\\g-2\end{pmatrix}$$

In order to make the 1st elements equal we require $k = \frac{1}{2}$,

Hence $$-2 = \tfrac{1}{2}(f-2) \quad \text{and} \quad -4 = \tfrac{1}{2}(g-2)$$

$f = -2$ and $g = -6$ and since $\mathbf{AB} = \frac{1}{2}\mathbf{BC}$ we have

$$\mathbf{AB} : \mathbf{BC} = 1 : 2$$

EXAMPLE 12

MNOPQRST is a parallelepiped in which MN = 1 unit, MP = 2 units and MQ = 3 units. Let **MN** = **a**, **MP** = **b** and **MQ** = **c** and the angles between each of these vectors, taken in pairs, be 60°. See Fig. 7.12.

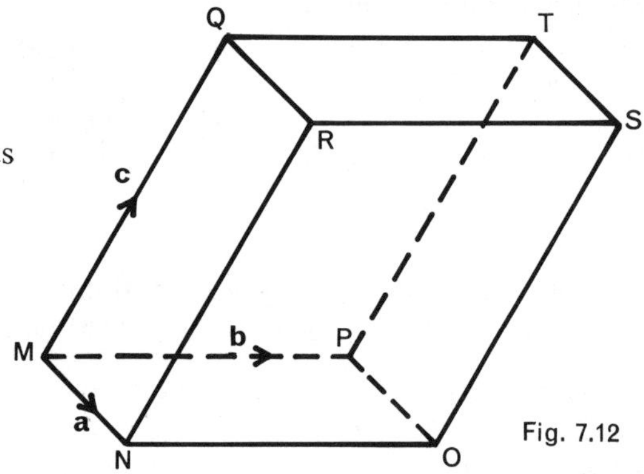

Fig. 7.12

(a) Calculate **a.b**, **a.c**, **b.c** and **a.a**.
(b) Calculate **(a+b+c).(a+b+c)** and hence write down the length of the space diagonal MS.
(c) Calculate **a.(a+b+c)** and then show that the cosine of the angle between MN and MS is 0.7.

(a) $\mathbf{a.b} = 1\times2\times\frac{1}{2} = 1 \qquad \mathbf{a.c} = 1\times3\times\frac{1}{2} = 1\frac{1}{2}$
$\mathbf{b.c} = 2\times3\times\frac{1}{2} = 3 \qquad \mathbf{a.a} = 1\times1\times1 = 1$

(b) $$\begin{aligned}(\mathbf{a}+\mathbf{b}+\mathbf{c}).(\mathbf{a}+\mathbf{b}+\mathbf{c}) &= \mathbf{a.a}+\mathbf{a.b}+\mathbf{a.c}+\mathbf{b.a}+\mathbf{b.b}+\mathbf{b.c}+\mathbf{c.a}+\mathbf{c.b}+\mathbf{c.c}\\ &= 1+1+1\tfrac{1}{2}+1+4+3+1\tfrac{1}{2}+3+9\\ &= 25\end{aligned}$$

But $(\mathbf{a}+\mathbf{b}+\mathbf{c}) = \mathbf{MS}$. Hence $|\mathbf{MS}| = \sqrt{25} = 5$ units.

(c) $\mathbf{a.(a+b+c)} = \mathbf{a.a}+\mathbf{a.b}+\mathbf{a.c} = 1+1+1\frac{1}{2} = 3\frac{1}{2}$

Also $\mathbf{a.(a+b+c)} = |\mathbf{a}||\mathbf{a}+\mathbf{b}+\mathbf{c}| \cos\theta = 1\times5\times\cos\theta$.

Hence $$5\cos\theta = 3\tfrac{1}{2}$$

$$\therefore \quad \cos\theta = 0.7$$

EXAMPLE 13

Three forces $\mathbf{F}_1 = 3\mathbf{i}-4\mathbf{j}+\mathbf{k}$, $\mathbf{F}_2 = 2\mathbf{i}+2\mathbf{j}-3\mathbf{k}$ and $\mathbf{F}_3 = \mathbf{i}+\mathbf{j}+4\mathbf{k}$ act at a point and cause the point to have a displacement represented by $\mathbf{d} = 14\mathbf{i}+2\mathbf{j}+2\mathbf{k}$. Calculate the work done by each force and the work done by the resultant force.

Work done by $\mathbf{F}_1 = \mathbf{F}_1.\mathbf{d} = (3\mathbf{i}-4\mathbf{j}+\mathbf{k}).(14\mathbf{i}+2\mathbf{j}+2\mathbf{k})$

$= 42-8+2$

$= 36$ units

Work done by $\mathbf{F}_2 = \mathbf{F}_2.\mathbf{d} = (2\mathbf{i}+2\mathbf{j}-3\mathbf{k}).(14\mathbf{i}+2\mathbf{j}+2\mathbf{k})$

$= 28+4-6$

$= 26$ units

Work done by $\mathbf{F}_3 = \mathbf{F}_3.\mathbf{d} = (\mathbf{i}+\mathbf{j}+4\mathbf{k}).(14\mathbf{i}+2\mathbf{j}+2\mathbf{k})$

$= 14+2+8$

$= 24$ units

Resultant force $\mathbf{F} = \mathbf{F}_1+\mathbf{F}_2+\mathbf{F}_3 = 6\mathbf{i}-\mathbf{j}+2\mathbf{k}$

Hence work done by $\mathbf{F} = \mathbf{F}.\mathbf{d} = (6\mathbf{i}-\mathbf{j}+2\mathbf{k}).(14\mathbf{i}+2\mathbf{j}+2\mathbf{k})$

$= 84-2+4$

$= 86$ units

Thus we see that the total work done by the individual forces, $(36+26+24)$ units, equals the work done by the resultant force.

Exercise 25

1) The vectors $\mathbf{p}$ and $\mathbf{q}$ have components $\begin{pmatrix}3\\2\end{pmatrix}$ and $\begin{pmatrix}2\\-1\end{pmatrix}$ respectively. Evaluate k and m if $k\mathbf{p}+m\mathbf{q} = \begin{pmatrix}4\\5\end{pmatrix}$.

2) The vectors $\mathbf{a}$, $\mathbf{b}$ and $\mathbf{c}$ have components $\begin{pmatrix}1\\-3\\2\end{pmatrix}$, $\begin{pmatrix}2\\1\\-2\end{pmatrix}$ and $\begin{pmatrix}3\\3\\4\end{pmatrix}$ respectively. Evaluate the components of the following vectors:

(a) $2\mathbf{a}+\mathbf{b}$ (b) $3\mathbf{a}-\mathbf{b}+\mathbf{c}$ (c) $4\mathbf{a}+2\mathbf{c}$ (d) $\mathbf{a}+2\mathbf{b}-3\mathbf{c}$

3) Find the magnitude of the vectors:

(a) $\mathbf{v} = \mathbf{i}+\mathbf{j}+\mathbf{k}$ (b) $\mathbf{r} = 3\mathbf{i}-4\mathbf{j}+\mathbf{k}$ (c) $\mathbf{u} = \frac{3}{5}\mathbf{i}-\frac{2}{5}\mathbf{j}+\frac{2\sqrt{3}}{5}\mathbf{k}$

4) Find real numbers x, y and z such that:

(a) $x\mathbf{i}+2y\mathbf{j}-z\mathbf{k}+3\mathbf{i}-\mathbf{j} = 4\mathbf{i}+3\mathbf{k}$ (b) $(x+4)\mathbf{i}+(y-5)\mathbf{j}+(z-1)\mathbf{k} = 0$

5) The work done by a force $\mathbf{F}$ causing displacement $\mathbf{d}$ is given by $\mathbf{F.d}$. Calculate the work done when $|\mathbf{F}| = 6$, $|\mathbf{d}| = 4$ and the angle between the vectors is 15°.

6) A ship sails on a bearing of 080° at 25 knots. The tide is running at 5 knots in the direction 045°. Find the resultant speed and direction of the ship.

7) A cylindrical bar of metal exerts a downward force of 200 N and is sitting in a 90° vee-block. Calculate the load on the sides of the block.

8) A steel girder, 3 m long, is exerting a downward force of 1000 N when suspended by a chain attached to either end. The height of the vee formed by the chain above the centre of the girder is 2 m. Calculate the tension in the chain.

9) Four forces acting on a point are represented in magnitude and direction by the vectors $3\mathbf{i}+2\mathbf{j}$, $2\mathbf{i}-7\mathbf{j}$, $-6\mathbf{i}+4\mathbf{j}$ and $2\mathbf{i}-\mathbf{j}$. Determine the resultant force vector, $\mathbf{F}$.

10) The resultant of two equal forces is equal to either of them. What is the angle between the two forces?

11) A(3, 5, 7), B(4, 4, 4) and C(5, −2, 4) are points such that $\mathbf{AB} = \mathbf{CD}$. Calculate the co-ordinates of point D.

12) The mid points of the sides of a triangle are (2, 4, 6), (3, 8, 7) and (5, 3, 2). Use vectors to find the co-ordinates of the vertices of the triangle.

13) The points P and Q have co-ordinates (2, 2, −2) and (6, 4, 2) with respect to an origin O.

(a) Calculate the co-ordinates of R which divides PQ in the ratio 3 : 1.
(b) Calculate the co-ordinates of S which divides PQ in the ratio 1 : 2.

14) A is the point (3, 4, 1), B is (9, 1, −5) and C is (11, 0, −7). Show that A, B and C are collinear and find the ratio in which B divides AC.

15) $\mathbf{a} = \begin{pmatrix} 1 \\ 4 \\ -2 \end{pmatrix}$, $\mathbf{b} = \begin{pmatrix} 6 \\ -2 \\ -1 \end{pmatrix}$ and $\mathbf{c} = \begin{pmatrix} 12 \\ -4 \\ -2 \end{pmatrix}$. Show that $\mathbf{a}$ is perpendicular to $\mathbf{b}$ and that $\mathbf{b}$ is parallel to $\mathbf{c}$.

16) Show that $\mathbf{d} = \frac{1}{5}\begin{pmatrix} 3 \\ 4 \end{pmatrix}$ is a unit vector. Calculate the magnitude of the resolved part of $\mathbf{a}$ in the direction of $\mathbf{d}$ given that:

(a) $\mathbf{a} = 2\mathbf{i}+3\mathbf{j}$ (b) $\mathbf{a} = 3\mathbf{i}-4\mathbf{j}$ (c) $\mathbf{a} = \mathbf{i}+\mathbf{j}$

17) The vertices of triangle ABC are A(3, 5, 0), B(1, 3, -1) and C(-1, 4, 1). Show that the triangle is right-angled isosceles using:

(a) The distance formula. (b) The scalar product.

18) P is the point (4, 3, -6), Q is (2, 1, -4) and R is (-1, 4, -1). Calculate the size of angle PQR.

19) A point 'moves' on a line with constant speed. At time $t = 0$ it is at (0, 0, 0). At the end of one second it is at (2, 4, $\sqrt{5}$). Write down the position vector to the point at the end of t seconds and state its speed in the x-direction, the y-direction and the z-direction.

20) A force of 39 N has two components of 16 N and 25 N. Show that the cosine of the angle between the components is $-\frac{4}{5}$.

21) Calculate the work done when a force $\mathbf{p} = 2\mathbf{i}+3\mathbf{j}+6\mathbf{k}$ moves its point of application through a displacement represented by $\mathbf{d} = 4\mathbf{i}-\mathbf{j}+2\mathbf{k}$.

22) Forces of magnitudes 9, 7 and 9 N act in the directions represented by $6\mathbf{i}-6\mathbf{j}+3\mathbf{k}$, $3\mathbf{i}+6\mathbf{j}-2\mathbf{k}$ and $7\mathbf{i}-4\mathbf{j}+4\mathbf{k}$ respectively. The three forces cause a displacement of a point by $8\mathbf{i}+8\mathbf{j}-4\mathbf{k}$ metres. Calculate the work done by each force and show that the total is equal to the work done by the resultant force.

23) P is the point (2, 3, -1), Q(0, 2, 1) and R(-4, 6, -3). Calculate the size of angle QPR and the area of triangle PQR.

24) $\mathbf{a} = \frac{1}{3}\begin{pmatrix} 2 \\ 2 \\ -1 \end{pmatrix}$ and $\mathbf{b} = \frac{1}{3}\begin{pmatrix} 2 \\ -1 \\ 2 \end{pmatrix}$. Show that **a** and **b** are perpendicular unit vectors and hence find vector **c**, of unit length, which is perpendicular to both **a** and **b**.

PROJECTS

1) An involute curve is formed when a line rolls on a circle without slipping. For example the end of a cotton thread being unwound from a reel, the thread being kept taut. Tables of the involute function have been compiled. If a circle rolls round the outside of another circle a point on the circumference describes a curve called an epicycloid. They can be produced from a set of Spirographs.

Find out the equations of one or both of these curves and, with the aid of drawings, show their relationship to the teeth of gear wheels.

2) Investigate the geometry required in the making of a particular form tool not discussed in this chapter. For example calculate the radius r required to form a radius R in the radial tool shown in Fig. 7.13.

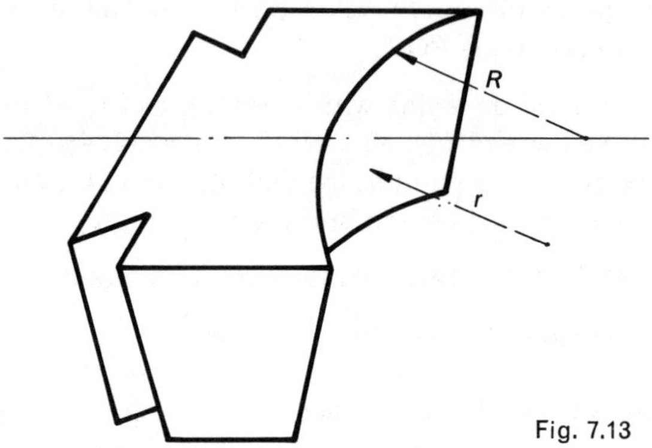

Fig. 7.13

3) Select a particular linked/sliding mechanism, for example a quick return motion, and with the aid of vector diagrams show the relative velocities of the ends of a rigid link or of two points in contact when sliding.

8. SAMPLING, ESTIMATION THEORY AND CHARTS

1. *Describe the concept of a sampling distribution.*
2. *Relate the mean and standard deviation of the sampling distribution to the mean and standard deviation of the parent population.*
3. *Recognise that the distribution for $\bar{x}$ is approximately normal when the sample size is not too small.*
4. *Calculate confidence interval estimates for the population mean using a sample from a population with a known standard deviation.*
5. *Determine unbiased estimates of the population mean and standard deviation from sample data.*
6. *Use the t-distribution to calculate the confidence interval for the population mean using a sample from a population with unknown standard deviation.*
7. *Calculate control limits and construct $\bar{x}$-charts and R-charts.*
8. *Obtain and use information from Gantt charts and Nomograms.*

SAMPLING DISTRIBUTIONS

Suppose that from a parent population we take a large number of samples each containing n items. We can calculate the mean for each of the samples and hence form a distribution for the sample means.

The samples which are chosen must be representative of the population. Representative samples may be obtained by *random sampling* in which each item of the population has an equal chance of being included in the sample. One way of obtaining a random sample is to give each item in the population a number and conduct a raffle. A second way is to use a table of *random numbers*, part of which is shown below, to obtain the sample.

RANDOM NUMBERS

50532	25496	95652	42657	73557	76152
50020	24819	52984	76168	07136	40876
79971	54195	25708	51817	36732	72484

EXAMPLE 1

The table below gives the diameters of 100 turned bars produced on an automatic lathe.

Diameter (mm)	Frequency
44.96	8
44.97	21
44.98	43
44.99	20
45.00	8

Three samples, each containing 5 items, are required from this population. Using a table of random numbers show how the samples are obtained.

Using two digits to number each of the 100 bars, the 8 bars of 44.96 mm diameter are numbered 00, 01, 02, 03, 04, 05, 06 and 07. The 21 bars of 44.97 mm diameter are numbered 08 to 28 and so on as shown in the table below.

Diameter (mm)	Frequency	Sampling number
44.96	8	00–07
44.97	21	08–28
44.98	43	29–71
44.99	20	72–91
45.00	8	92–99

Sampling numbers are now taken from the table of random numbers given above. From the first line the sequence 50, 53, 22, 54 and 96 is obtained. Hence the first sample will comprise bars numbered 50, 53, 22, 54 and 96. Carrying on in this way we find the second sample to comprise bars numbered 95, 65, 24, 26 and 57. The third sample comprises bars numbered 73, 55, 77, 61 and 52.

A population may be finite or it may be infinite. If we draw 5 balls successively from a box containing 100 balls, without replacing them, we are sampling from a finite population. If, however, we replace each ball after drawing it we are sampling from an infinite population, since theoretically we may draw any number of samples without exhausting the population.

In practice, sampling without replacement from a very large population may be considered as sampling from an infinite population.

DISTRIBUTION OF SAMPLE MEANS

If the population consists of n_p items it can be shown that:

$$\bar{x} = \mu$$

and

$$s_{\bar{x}} = \frac{\sigma}{\sqrt{n}}\sqrt{\frac{n_p - n}{n_p - 1}}$$

where μ = the mean of the population

$\bar{x}$ = the mean of the sampling distribution

σ = the standard deviation of the population

$s_{\bar{x}}$ = the standard deviation of the sampling distribution

and n = the number of items in each sample

If the population is infinite, or the sampling is with replacement then the results given above reduce to:

$$\bar{x} = \mu \quad \text{and} \quad s_{\bar{x}} = \frac{\sigma}{\sqrt{n}}$$

For values of n equal to, or greater than 30, the distribution for the sample means is approximately a normal distribution with a mean of $\bar{x}$ and standard deviation of $s_{\bar{x}}$ irrespective of whether the population itself is a normal distribution. The approximation to the normal distribution gets better as n gets larger.

If the population is normally distributed and σ is known the distribution for sample means is normally distributed even for small values of n (Fig. 8.1).

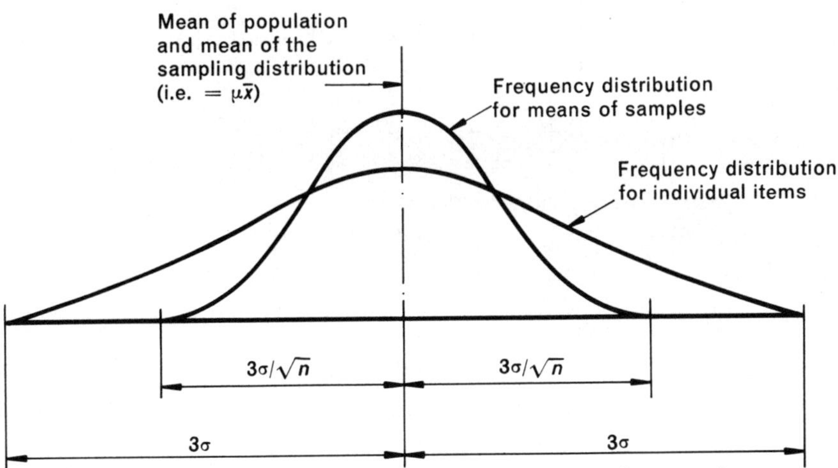

The frequency curve for the sample means is a normal curve whose standard deviation is $\sigma/\sqrt{n}$.

Fig. 8.1

STANDARD ERROR OF THE DISTRIBUTION OF SAMPLE MEANS

The standard deviation of a sampling distribution statistic is called the **standard error.** Thus for the distribution of sample means the standard error is $s_{\bar{x}} = \dfrac{\sigma}{\sqrt{n}}$.

EXAMPLE 2

A population consists of the five numbers 1, 2, 3, 4 and 5. Find:

(a) the mean and standard deviation of the population,

(b) the mean and standard deviation of all the possible samples of size 2 which can be drawn with replacement from this population.

(a) The mean of the population is:

$$\mu = \frac{1+2+3+4+5}{5} = \frac{15}{5} = 3$$

The standard deviation of the population is:

$$\begin{aligned}\sigma &= \sqrt{\frac{\Sigma x^2}{N} - \mu^2} \\ &= \sqrt{\frac{1^2+2^2+3^2+4^2+5^2}{5} - 3^2} \\ &= \sqrt{2} = 1.414\end{aligned}$$

(b) There are 25 samples of size 2 which can be drawn, with replacement, from the population. They are:

1, 1	2, 1	3, 1	4, 1	5, 1
1, 2	2, 2	3, 2	4, 2	5, 2
1, 3	2, 3	3, 3	4, 3	5, 3
1, 4	2, 4	3, 4	4, 4	5, 4
1, 5	2, 5	3, 5	4, 5	5, 5

The corresponding sample means are:

1.0	1.5	2.0	2.5	3.0
1.5	2.0	2.5	3.0	3.5
2.0	2.5	3.0	3.5	4.0
2.5	3.0	3.5	4.0	4.5
3.0	3.5	4.0	4.5	5.0

The mean of the sampling distribution is:

$$\bar{x} = \frac{\text{sum of all the sample means}}{\text{number of samples}} = \frac{75}{25} = 3$$

This illustrates the fact that:

$$\bar{x} = \mu$$

The standard deviation of the sampling distribution is:

$$s_{\bar{x}} = \sqrt{\frac{\Sigma \bar{x}_i^2}{N} - \bar{x}^2}$$

where $\bar{x}_i$ = the means of the ith sample

and N = the number of samples taken

$$s_{\bar{x}} = \sqrt{\frac{250}{25} - 3^2} = 1$$

This illustrates the fact that for sampling with replacement:

$$s_{\bar{x}} = \frac{\sigma}{\sqrt{n}} = \frac{\sqrt{2}}{\sqrt{2}} = 1$$

EXAMPLE 3

The heights of 2000 male factory workers are normally distributed with a mean of 172 cm and a standard deviation of 7.5 cm. If 40 samples each containing 30 workers are taken, determine the mean and standard deviation of the sampling distribution for sample means if:

(a) the sampling is done with replacement,
(b) the sampling is done without replacement.

(a) $\bar{x} = \mu = 172$ cm

$$s_{\bar{x}} = \frac{\sigma}{\sqrt{n}} = \frac{7.5}{\sqrt{30}} = 1.369 \text{ cm}$$

(b) $\bar{x} = \mu = 172$ cm

$$s_{\bar{x}} = \frac{\sigma}{\sqrt{n}}\sqrt{\frac{n_p - n}{n_p - 1}} = \frac{7.5}{\sqrt{30}}\sqrt{\frac{2000-30}{2000-1}} = 1.359 \text{ cm}$$

Thus the value of $s_{\bar{x}}$ for sampling done without replacement is only slightly less than that for sampling done with replacement. Hence the population of 2000 is large enough to be considered an infinite population.

EXAMPLE 4

A very large batch of ball bearings has a mean diameter of 25.02 mm with a standard deviation of 0.30 mm. A random sample of 100 of these ball bearings is taken. Find:

(a) the probability that the mean of the sample lies between 24.96 mm and 25.00 mm,
(b) the probability that the mean of the sample is greater than 25.10 mm.

For the sampling distribution of means

$$\bar{x} = \mu = 25.02 \text{ mm}$$

$$s_{\bar{x}} = \frac{\sigma}{\sqrt{n}} = \frac{0.30}{\sqrt{100}} = 0.03 \text{ mm}$$

(a) When $\bar{x} = 24.96$ mm, $u_1 = \dfrac{\bar{x}-\mu}{s_{\bar{x}}} = \dfrac{24.96-25.02}{0.03} = -2.00.$

When $\bar{x} = 25.00$ mm, $u_2 = \dfrac{\bar{x}-\mu}{s_{\bar{x}}} = \dfrac{25.00-25.02}{0.03} = -0.67.$

Since $n > 30$ we may use the normal distribution to determine probabilities. From a Table of Areas and using Fig. 8.2

area between $u = -2.00$ and $u = 0 = 0.477\,2$

area between $u = -0.67$ and $u = 0 = 0.248\,6$

area between $u = -2.0$ and $u = -0.67 = 0.477\,2-0.248\,6$

$= 0.228\,6$

Hence the probability that the mean of the sample lies between 24.96 mm and 25.00 mm is 0.228 6.

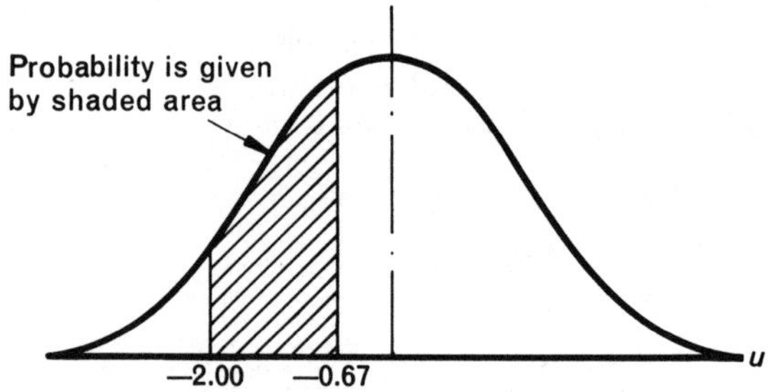

Fig. 8.2

(b) When $\bar{x} = 25.10$, $u = \dfrac{25.10-25.02}{0.03} = 2.67.$

From a Table of Areas and using Fig. 8.3

area between $u = 0$ and $u = 2.67 = 0.496\,2$

tail area beyond $u = 2.67 = 0.5-0.496\,2 = 0.003\,8$

Hence the probability that the mean of the sample is greater than 25.10 mm is 0.003 8.

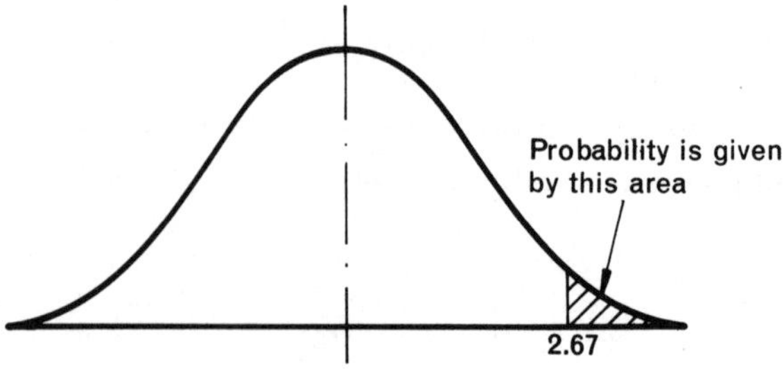

Fig. 8.3

CONFIDENCE LIMITS

It is seldom possible to measure every item in a population to find the mean of that population. Hence we must resort to sampling. To estimate the mean of a population we take a sample from the population and calculate the mean of the sample. As we have previously seen, we then regard the mean of the sample as an estimate of the mean of the population. This estimate is called a **point estimate** of the mean of the population because it is a single number.

To determine the reliability of the point estimate of the population mean we use confidence limits. We know, from Tables, that for a normal distribution of sample means only 5% of these sample means are likely to lie outside the limits of $\mu \pm \frac{1.96\sigma}{\sqrt{n}}$ (see Fig. 8.4). We say that at the 95% confidence level the true mean of the population is unlikely to lie outside the limits of $\bar{x} \pm \frac{1.96\sigma}{\sqrt{n}}$, $\bar{x}$ being the mean of the sample.

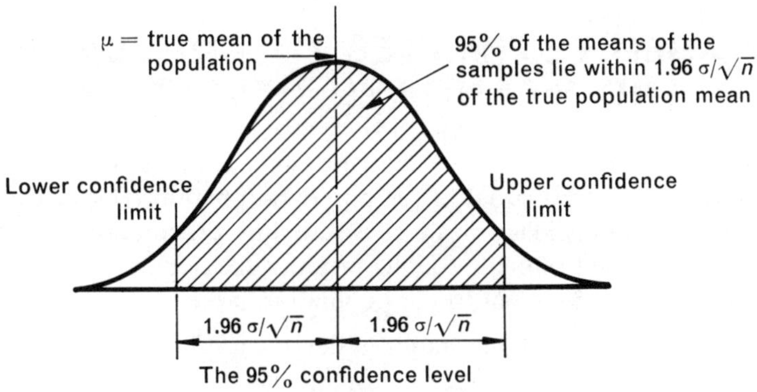

Fig. 8.4

The most commonly used confidence levels are the 95%, 99% and 99.9% levels. The table below shows the limits for these confidence levels.

Confidence level	95%	99%	99.9%
Limits	$\bar{x} \pm \frac{1.96\sigma}{\sqrt{n}}$	$\bar{x} \pm \frac{2.58\sigma}{\sqrt{n}}$	$\bar{x} \pm \frac{3.29\sigma}{\sqrt{n}}$

These confidence limits are examples of *interval estimates* of the population mean because we are stating limits between which the population mean may be expected to lie. Although the sample mean is useful as an estimate of the population mean there is no way by which we can express the accuracy of the estimate. (In fact, mathematically speaking, the probability of the mean of the sample exactly equalling the mean of the population is $p = 0$.) However, by using confidence limits, the degree of accuracy of the estimate is given by the confidence level.

EXAMPLE 5

The diameters of a random sample of 100 turned bars were measured and their mean diameter was found to be 18.10 mm. The standard deviation for the population is known to be 1.2 mm. At the 99% confidence level, what is the estimate of the mean diameter of all the turned bars produced?

From the table above, the limits for the 99% confidence level are

$$\bar{x} \pm \frac{2.58\sigma}{\sqrt{n}} = 18.10 \pm \frac{2.58 \times 1.2}{\sqrt{100}} = 18.10 \pm 0.31 \text{ mm}$$

We can be 99% certain that the true mean of the population lies between 17.79 mm and 18.41 mm.

UNBIASED ESTIMATES OF POPULATION MEAN AND STANDARD DEVIATION

An *unbiased estimator* is a sample statistic (such as the mean of a sample) whose expected value is equal to the population parameter (for example, the population mean). The mean of a sampling distribution for sample means is an unbiased estimate for the mean of the population because its value is expected to be equal to the population mean.

The unbiased estimate for the standard deviation of a population is:

$$\hat{s} = s\sqrt{\frac{n-1}{n}}$$

where $\hat{s}$ = the unbiased estimate of the population standard deviation (i.e. $\hat{s} = \sigma$).

s = standard deviation of the sample

n = the sample size

EXAMPLE 6

A sample of 5 resistors were checked with the following results: 12.6, 12.7, 12.5, 12.6 and 12.8 ohm. Determine the unbiased estimates of the mean and standard deviation of the population from which the resistors were drawn.

$$\bar{x} = \frac{12.6+12.7+12.5+12.6+12.8}{5} = \frac{63.2}{5} = 12.64 \text{ ohm}$$

This is an unbiased estimate of the population mean.

$$s = \sqrt{\frac{\Sigma x^2}{n} - \bar{x}^2} = \sqrt{\frac{798.9}{5} - 12.64^2}$$

$$= \sqrt{0.0104} \quad = 0.102 \text{ mm}$$

This is an estimate of the standard deviation of the population but it is not unbiased.

The unbiased estimate of the population standard deviation is

$$\hat{s} = \sqrt{\frac{n-1}{n}} s = 0.102 \times \sqrt{\frac{5-1}{5}} = 0.081\,6 \text{ ohm}$$

(Note that for large samples $(n - 30)$ the correction $\sqrt{\frac{n-1}{n}}$ is insignificant. Hence for large samples the unbiased estimate of σ may be taken to be s.)

THE *t*-DISTRIBUTION

On page 113 it was stated that the normal distribution may be used to estimate a population mean for any large sample $(n \geqslant 30)$ and for a small sample $(n < 30)$ only if the population is normally distributed and σ is known.

In this section we consider the situation in which the sample is small and the population is normally distributed but σ is not known.

Let us define t by the equation:

$$t = \frac{\bar{x} - \mu}{\hat{s}/\sqrt{n}}$$

which is analogous to:

$$u = \frac{\bar{x} - \mu}{s_{\bar{x}}} = \frac{\bar{x} - \mu}{\sigma/\sqrt{n}} \quad \text{(see page 116)}$$

Now $$\hat{s} = s\sqrt{\frac{n}{n-1}}$$

$\therefore$ $$t = \frac{|\bar{x} - \mu|\sqrt{n-1}}{s}$$

This latter formula for t includes s in the denominator. Now s, the standard deviation of the sample, is not a constant as is σ, but varies with each sample mean. The result is that the values of t are not distributed in the same way as the values of u.

From a normal distribution (i.e. a population with a normal distribution curve) whose standard deviation is not known, let us take a number of samples of size n. For each sample let us calculate the value of t using the mean of the sample, $\bar{x}$, and the standard deviation of the sample, s, as shown in Example 7.

EXAMPLE 7

A sample of 5 items is taken from a population which is normally distributed. If the mean of the population is 17.2 mm and the mean and standard deviation are 18.0 mm and 0.5 mm respectively, calculate the value of t.

$$\hat{s} = s\sqrt{\frac{n}{n-1}} = 0.5 \times \sqrt{\frac{5}{5-1}} = 0.56 \text{ mm}$$

$$t = \frac{|\bar{x}-\mu|}{\hat{s}/\sqrt{n}} = \frac{|18.0-17.2|}{0.56/\sqrt{5}} = \frac{0.8\times\sqrt{5}}{0.56} = 3.19$$

From these calculated values of t a distribution curve is obtained and the distribution is called a ***t*-distribution**. By varying the sample size a series of distribution curves similar to those shown in Fig. 8.5 are obtained. The curves show that when the sample is large $(n \geqslant 30)$ the t-distribution is very nearly normal. However, when the samples are small the t-distribution is no longer normal and its approximation to the normal distribution gets worse as the sample size decreases.

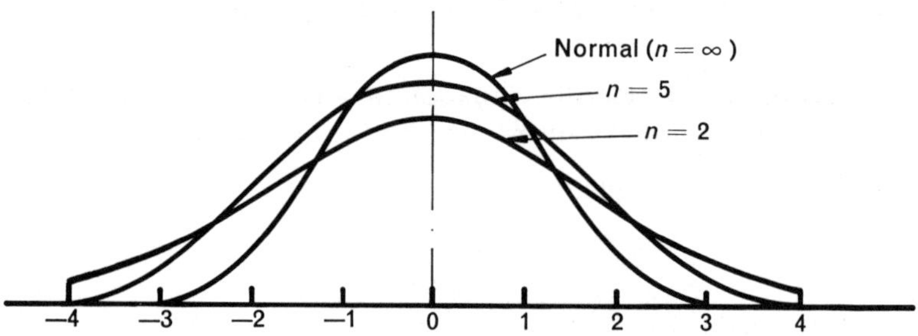

Fig. 8.5

This means that when the sample is large $(n \geqslant 30)$ we may use the normal distribution to determine confidence limits (see page 117). However, when the sample is small $(n < 30)$ the t-distribution must be used to determine the confidence limits.

Table 8.1 shows the t-distribution in which values of t are tabulated against degrees of freedom. The number of degrees of freedom for a single sample is:

$$\text{d.f.} = n-1$$

that is, the number of degrees of freedom is one less than the number of items in the sample.

CONFIDENCE LIMITS FOR SMALL SAMPLES

As was done with the normal distribution we can define the 95% and the 99% confidence limits by using the table of the t-distribution (Table 8.1).

TABLE 8.1 The t-distribution

Significance level	5%		1%	
	One-tailed test	Two-tailed test	One-tailed test	Two-tailed test
1	6.314	12.706	31.821	63.657
2	2.920	4.303	6.965	9.925
3	2.353	3.182	4.541	5.841
4	2.132	2.776	3.747	4.604
5	2.015	2.571	3.365	4.032
6	1.943	2.447	3.143	3.707
7	1.895	2.365	2.998	3.499
8	1.860	2.306	2.896	3.355
9	1.833	2.262	2.821	3.250
10	1.812	2.228	2.764	3.169
11	1.796	2.201	2.718	3.106
12	1.782	2.179	2.681	3.055
13	1.771	2.160	2.650	3.012
14	1.761	2.145	2.624	2.977
15	1.753	2.131	2.602	2.947
16	1.746	2.120	2.583	2.921
17	1.740	2.110	2.567	2.898
18	1.734	2.101	2.552	2.878
19	1.729	2.093	2.539	2.861
20	1.725	2.086	2.528	2.845
21	1.721	2.080	2.518	2.831
22	1.717	2.074	2.508	2.819
23	1.714	2.069	2.500	2.807
24	1.711	2.064	2.492	2.797
25	1.708	2.060	2.485	2.787
26	1.706	2.056	2.479	2.779
27	1.703	2.052	2.473	2.771
28	1.701	2.048	2.467	2.763
29	1.699	2.045	2.462	2.756
30	1.697	2.042	2.457	2.750

We can represent the confidence limits for a population mean by:

$$\bar{x} \pm t_c \frac{\sigma}{\sqrt{n}}$$

where t_c is the value of t obtained from Table 8.1.

Fig. 8.6 shows how the 95% confidence limits are obtained.

Since both tails of the t-distribution curve are used we look up the value of t_c in the 'two-tailed test' column.

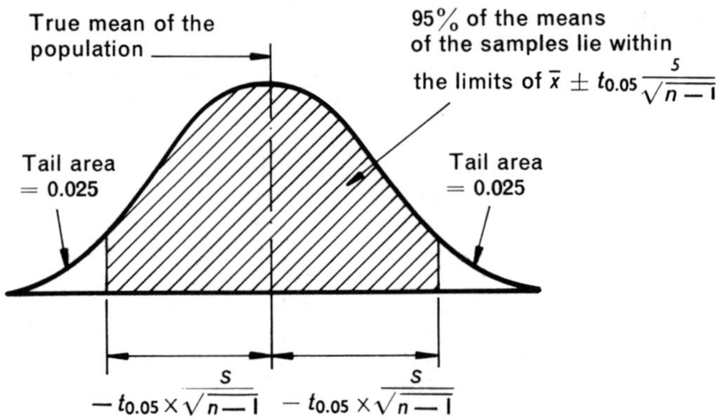

Fig. 8.6

Since σ, the standard deviation of the population is generally unknown we use the unbiased estimate for σ which is:

$$\hat{s} = s\sqrt{\frac{n}{n-1}}$$

Hence the confidence limits for a population mean is given by:

$$\bar{x} \pm t_c \times \sqrt{\frac{n}{n-1}} \times s \times \frac{1}{\sqrt{n}} = \bar{x} \pm \frac{t_c s}{\sqrt{n-1}}$$

where $\bar{x}$ = the mean of the sample

t_c = the value of t for a two-tailed test

s = standard deviation of the sample

n = number of items in the sample

EXAMPLE 8

The diameters of a random sample of 10 ball bearings were measured and their mean diameter was found to be 4.38 mm with a standard deviation of 0.06 mm. Find:

(a) the 95%,
(b) the 99% confidence limits for the population mean from which the sample was taken.

(a) The 95% confidence limits are given by:

$$\bar{x} \pm t_{0.05} \times \frac{s}{\sqrt{n-1}}$$

Since d.f. $= 10-1 = 9$, $t_{0.05} = 2.262$.

Also $\bar{x} = 4.38$ mm, $s = 0.06$ mm and $n = 10$.

$$\therefore \quad \bar{x} \pm t_{0.05} \times \frac{s}{\sqrt{n-1}} = 4.38 \pm 2.262 \times \frac{0.06}{\sqrt{9}}$$

$$= 4.38 \pm 0.045$$

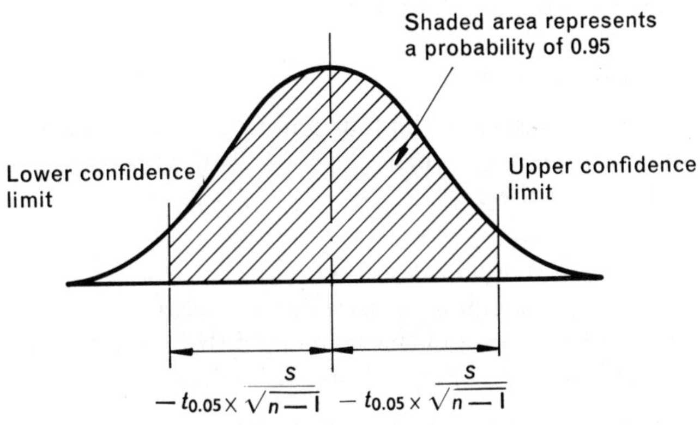

Fig. 8.7

Hence we can be 95% certain that the mean of the population of ball bearings lies between 4.335 mm and 4.425 mm (see Fig. 8.7).

(b) The 99% confidence limits are given by

$$\bar{x} \pm t_{0.01} \times \frac{s}{\sqrt{n-1}}$$

For d.f. $= 9$, $t_{0.01} = 3.250$.

$$\therefore \quad \bar{x} \pm t_{0.01} \times \frac{s}{\sqrt{n-1}} = 4.38 \pm 3.250 \times \frac{0.06}{\sqrt{9}}$$

$$= 4.38 \pm 0.065$$

We can be 99% certain that the mean of the population of ball bearings lies between 4.315 mm and 4.445 mm. (Note that in obtaining the confidence limits a 'two-tailed' value for t is used.)

Exercise 26

1) The diameters of 5000 electric motor shafts are normally distributed with a mean of 50.42 mm and a standard deviation of 1.23 mm. If 80 samples each containing 30 shafts are obtained find the expected mean and standard deviation of the sampling distribution for sample means if the sampling was done:

(a) with replacement,
(b) without replacement.
(c) How many samples are likely to have a mean between 50.32 mm and 50.48 mm?

2) 400 resistors have a mean resistance of 50.2 ohm and a standard deviation of 0.5 ohm. Find the probability that a random sample of 50 of these resistors would have a resistance:

(a) between 50.2 ohm and 50.3 ohm,
(b) less than 50.05 ohm,
(c) greater than 50.18 ohm.

3) A population consists of the four numbers 2, 6, 10 and 14. Consider all the samples of size two which can be drawn, with replacement, from this population. Calculate:

(a) the population mean,
(b) the population standard deviation,
(c) the mean of the sampling distribution for means,
(d) the standard deviation of the sampling distribution for means.
(e) Verify (c) and (d) by using suitable formulae.

4) The masses of 3000 ball bearings are normally distributed with a mean of 635 grams and a standard deviation of 1.4 grams. If 70 samples each containing 36 ball bearings are taken from this population, find the expected mean and standard deviation of the distribution of samples if the sampling is done:

(a) with replacement,
(b) without replacement.

5) A sample of 5 components produced on a machine tool were measured. Their lengths were found to be 18.99, 19.11, 19.08, 18.96 and 19.10 mm. Determine the unbiased estimates of:

(a) the mean of the large batch from which this sample was drawn,
(b) the standard deviation of the batch.

6) The heights of 100 male factory workers represent a random sample of the heights of all 1972 male workers in the factory. The mean height of the 100 men measured was 1.727 m with a standard deviation of 0.05 m. Determine unbiased estimates of:

(a) the mean height of the population,
(b) the standard deviation of the population.

7) A random sample is drawn from a population with a known standard deviation of 2.0.

(a) Calculate the standard error of the sample mean if the sample size is:
(i) 9 (ii) 25 (iii) 100
(b) If the sample mean is 25.3, calculate the 95% confidence limits if the sample size is:
(i) 9 (ii) 25 (iii) 100

8) The percentage of zinc in a certain alloy is measured six times. The standard deviation of these measurements is 2.5% and the mean of the six measurements is 14.1%. Determine the 95% confidence limits for the true percentage of zinc assuming:

(a) that the measurements are normally distributed,
(b) that the t-distribution is applicable.

9) The diameter of a sphere is measured 10 times. The mean of these measurements is found to be 53.6 mm with a standard deviation of 0.8 mm. Find:

(a) 95%,
(b) 99% confidence limits for the true diameter of the sphere using the t-distribution.

10) If in Question 9, the measurements were normally distributed, determine

(a) 95%,
(b) 99% confidence limits

for the true diameter of the sphere.

STATISTICAL CHARTS

Control Charts

Quality control is the application of statistical principles to mass production processes with the aims of ensuring uniformity in the articles produced. Any manufacturing process inevitably has a variability which cannot be completely eliminated. Control charts are used to help contain this variability by giving a warning when the variation is due to assignable causes such as faulty machine settings, worn parts, poor operators or raw materials of low quality.

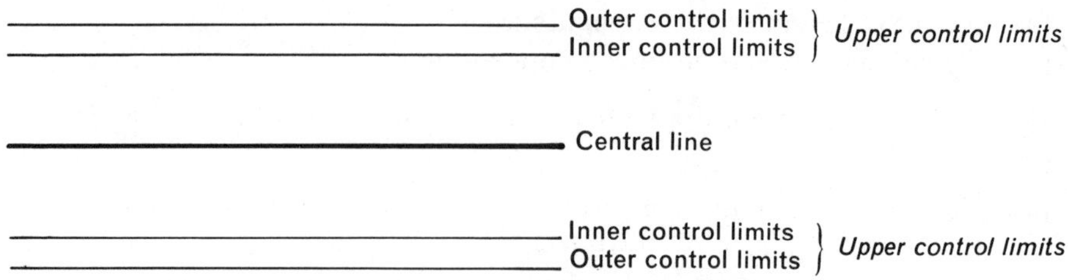

Fig. 8.8

A simple form of control chart consists of a **central line** corresponding to an 'average quality' and two pairs of **control limits**. See Fig. 8.8.

'Average quality' refers to some measurable property such as the breaking strength of a cord, the diameter of a rod or the number of defective screws in a sample. The upper and lower control limits are chosen so that the variation of values between them is attributed to chance, whereas values outside the limits are indicative of lack of control.

Average quality is controlled by plotting the means of successive samples in time sequence on a **control chart for means** or $\bar{x}$-chart. Variability is controlled by plotting the sample ranges on an R-chart or sample standard deviations on an S-chart.

British practice favours the use of control limits corresponding to the specific levels shown in Fig. 8.9 which means that we can expect 1 sample mean in 20 to fall outside the inner control limits and 1 in 500 outside the outer limits. (American practice favours $\mu \pm 3\dfrac{\sigma}{\sqrt{n}}$.)

The outer limits are called **action** limits, since if a point falls outside them it is almost certain that the process has altered and some action is necessary to restore control. The inner limits are called **warning** limits. No action is necessary if one sample point falls outside them but if two successive points are outside then some action should be taken.

If the production has been measured over a long period of time then μ and σ may be known and the limits in Fig. 8.9 can be calculated. It is more

Fig. 8.9

usual, however, that estimates will have to be made. As samples greater than 30 are unlikely, μ is estimated from $\bar{x}$, the mean of means of the small samples, and σ from the *mean range* of this set of small samples. Since the range is always greater than the standard deviation we must reduce it by multiplying by a number, a_n, where n is the sample size.

Thus $\mu+1.96\frac{\sigma}{\sqrt{n}}$ is estimated from $\bar{x}+1.96\frac{a_n\bar{R}}{\sqrt{n}}$ and if the values for $1.96\frac{a_n}{\sqrt{n}}$ are replaced by $A'_{0.025}\,\bar{R}$ we have $\mu+1.96\frac{\sigma}{\sqrt{n}} = \bar{x}+A'_{0.025}\,\bar{R}$.

The values of a_n, $A'_{0.001}$ and $A'_{0.025}$ for sample sizes of $n = 2, 3, 4, \ldots, 10$ are given in Table 8.2.

This estimation for σ should only be used for sample sizes $\leqslant 10$ since the estimate becomes unreliable for larger samples.

TABLE 8.2

n	2	3	4	5	6	7	8	9	10
a_n	0.886 2	0.590 8	0.485 7	0.429 9	0.394 6	0.369 8	0.351 2	0.336 7	0.324 9
$A'_{0.001}$	1.94	1.05	0.75	0.59	0.50	0.43	0.38	0.35	0.32
$A'_{0.025}$	1.23	0.67	0.48	0.37	0.32	0.27	0.24	0.22	0.20

Hence for small samples the lines of the control chart are given by Fig. 8.10, where $\bar{x}$ is the mean of the means of the small samples and $\bar{R}$ is the mean range.

Fig. 8.10

EXAMPLE 9

A small transmitter is designed to generate and transmit an automatic signal. The specification requires an average signal time of 30.0 seconds and standard deviation of 0.8 seconds. Calculate the control limits for means of samples of 5 transmitters from the production line.

To determine the control limits for an $\bar{x}$-chart using $\mu = 30.0$ seconds and $\sigma = 0.8$ seconds we have:

Upper control limits $\begin{cases} \text{Outer} = \mu + 3.09\dfrac{\sigma}{\sqrt{n}} = 30.0 + 3.09\dfrac{0.8}{\sqrt{5}} = 31.1 \text{ seconds} \\ \text{Inner} = \mu + 1.96\dfrac{\sigma}{\sqrt{n}} = 30.0 + 1.96\dfrac{0.8}{\sqrt{5}} = 30.7 \text{ seconds} \end{cases}$

Lower control limits $\begin{cases} \text{Inner} = \mu - 1.96\dfrac{\sigma}{\sqrt{n}} = 30.0 - 1.96\dfrac{0.8}{\sqrt{5}} = 29.3 \text{ seconds} \\ \text{Outer} = \mu - 3.09\dfrac{\sigma}{\sqrt{n}} = 30.0 - 3.09\dfrac{0.8}{\sqrt{5}} = 28.9 \text{ seconds} \end{cases}$

EXAMPLE 10

A technician measures the hydrogen ion concentration from samples of size four taken at hourly intervals. Construct an $\bar{x}$-chart on the basis of the data tabulated below for 25 preliminary samples and check whether these data are acceptable for the purpose of calculating control limits. (See Table 8.3.)

$$\bar{x} = \frac{178.825}{25} = 7.153$$

$$\bar{R} = \frac{12.0}{25} = 0.48$$

Hence for an $\bar{x}$-chart the Central Line takes the value $\bar{x} = 7.153$

$$\text{Outer} = \bar{x} + A'_{0.001}\,\bar{R} = 7.153 + 0.75(0.48) = 7.513$$
$$\text{Inner} = \bar{x} + A'_{0.025}\,\bar{R} = 7.153 + 0.48(0.48) = 7.383$$

$$\text{Inner} = \bar{x} - A'_{0.025}\,\bar{R} = 7.153 - 0.48(0.48) = 6.923$$
$$\text{Outer} = \bar{x} - A'_{0.001}\,\bar{R} = 7.153 - 0.75(0.48) = 6.793$$

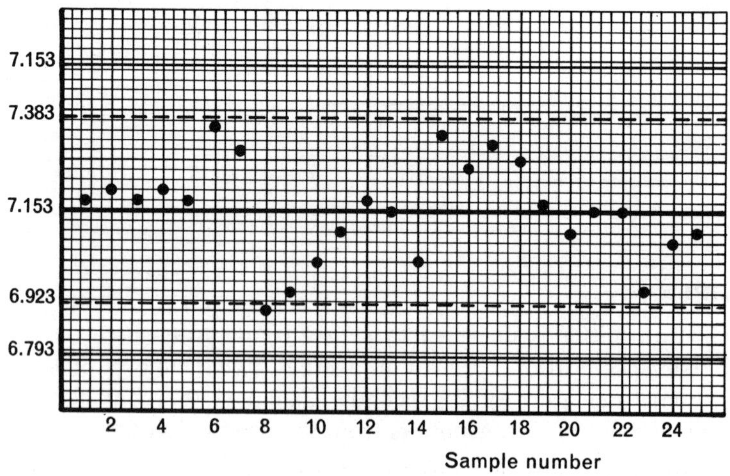

Fig. 8.11

TABLE 8.3

Sample No.	Sample measurements (pH)				$\bar{x}$	R
1	7.0	7.0	7.4	7.3	7.175	0.4
2	7.0	6.9	7.6	7.3	7.200	0.7
3	7.3	7.0	7.0	7.4	7.175	0.4
4	7.3	7.1	7.4	7.0	7.200	0.4
5	7.3	7.4	7.0	7.0	7.175	0.4
6	7.5	7.2	7.5	7.3	7.375	0.3
7	7.5	7.4	7.1	7.2	7.300	0.4
8	7.0	6.5	7.2	6.9	6.900	0.7
9	7.0	7.2	6.5	7.1	6.950	0.7
10	7.0	6.6	7.4	7.1	7.025	0.8
11	7.1	7.4	7.0	6.9	7.100	0.5
12	7.2	7.4	6.9	7.2	7.175	0.5
13	7.0	7.0	7.3	7.3	7.150	0.3
14	7.0	6.9	7.0	7.2	7.025	0.3
15	7.4	7.6	7.0	7.4	7.350	0.6
16	7.3	7.3	7.4	7.0	7.250	0.4
17	7.3	7.5	7.5	7.0	7.325	0.5
18	7.4	7.2	7.4	7.1	7.275	0.3
19	7.0	7.4	7.1	7.2	7.175	0.4
20	7.0	7.3	7.2	6.9	7.100	0.4
21	7.0	7.4	7.0	7.2	7.150	0.4
22	7.2	6.8	7.3	7.3	7.150	0.5
23	6.5	7.4	7.0	6.9	6.950	0.9
24	7.1	7.1	6.9	7.2	7.075	0.3
25	6.9	7.0	7.1	7.4	7.100	0.5
				Total	178.825	12.0

The $\bar{x}$-chart is shown in Fig. 8.11 with the points for the preliminary data plotted. It can be seen that the process is in control and that the preliminary data are acceptable for the purpose of calculating control limits.

While 'average quality' is controlled by plotting the means of successive samples on an $\bar{x}$-chart, *variability* is controlled by plotting the sample ranges on an R-chart, where, using British practice:

$$\text{Central line} : \bar{R}$$

$$\textit{Upper control limits}\begin{cases}\text{Outer} = D'_{0.999}\,\bar{R}\\ \text{Inner} = D'_{0.975}\,\bar{R}\end{cases}$$

$$\textit{Lower control limits}\begin{cases}\text{Inner} = D'_{0.025}\,\bar{R}\\ \text{Outer} = D'_{0.001}\,\bar{R}\end{cases}$$

The values of the D''s are given in Table 8.4

TABLE 8.4

n	2	3	4	5	6	7	8	9	10
$D'_{0.001}$	0.00	0.04	0.10	0.16	0.21	0.26	0.29	0.32	0.35
$D'_{0.025}$	0.04	0.18	0.29	0.37	0.42	0.46	0.50	0.52	0.54
$D'_{0.975}$	2.81	2.17	1.93	1.81	1.72	1.66	1.62	1.58	1.56
$D'_{0.999}$	4.12	2.98	2.57	2.34	2.21	2.11	2.04	1.99	1.94

Lower limits are usually ignored in $\bar{R}$-charts since we are primarily concerned with ensuring that the process variability does *not increase.* If points fall below the lower control limits a decrease in variability is indicated and since this is, in general, a desirable event, an attempt should be made to determine and maintain this improvement in precision.

EXAMPLE 11

Use the information given in Table 8.3 to construct an $\bar{R}$-chart for the results of the 25 small samples in Example 10.

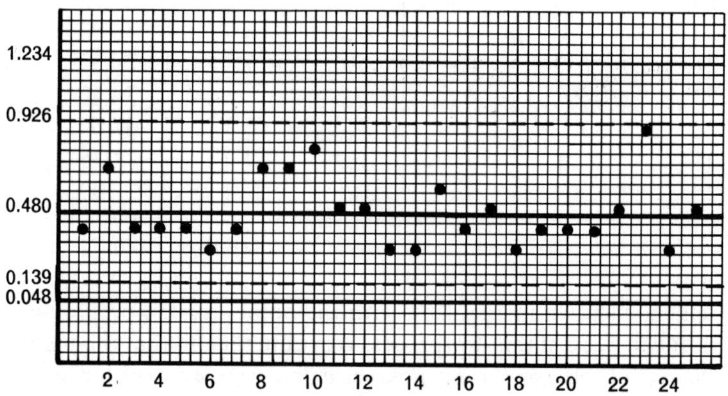

Fig. 8.12

$$\text{Central line : } \bar{R} = 0.48$$

$$\textit{Upper control limits}\begin{cases}\text{Outer} = D'_{0.999}\,\bar{R} = 2.57(0.48) = 1.234\\ \text{Inner} = D'_{0.975}\,\bar{R} = 1.93(0.48) = 0.926\end{cases}$$

$$\textit{Lower control limits}\begin{cases}\text{Inner} = D'_{0.025}\,\bar{R} = 0.29(0.48) = 0.139\\ \text{Outer} = D'_{0.001}\,\bar{R} = 0.10(0.48) = 0.048\end{cases}$$

Gantt Progress Charts

There are many types of Gantt charts but all are special bar diagrams scaled in units of time. The one shown in Fig. 8.13 is a progress chart where actual performance is compared with planned performance.

Although each equal length in Fig. 8.13 represents an equal period of time, in this case one week, it does not necessarily represent the same planned production. The first figure in week 1 (150) is the planned production number and the second (150) is the planned total. The first figure in the second week (200) is the planned production for that week and the second figure (350) is the planned cumulative total; and so on.

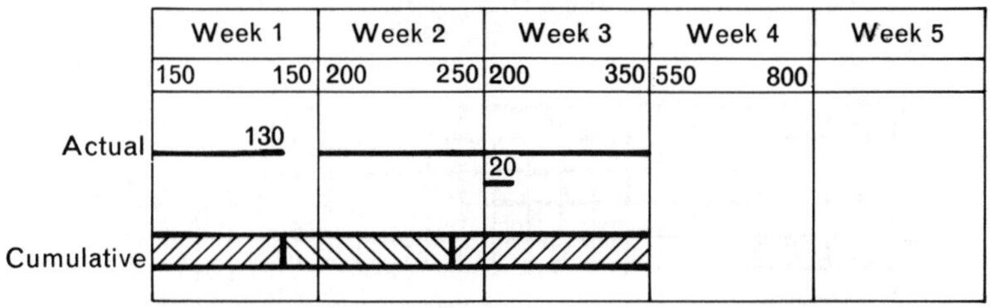

Fig. 8.13

The thin line represents the *actual* production and the thick line the *cumulative total.* Hence week 3 shows a planned production of 200 units but the thin line indicates an actual production of 220 which makes up for the shortfall in week 1 keeping the production up to schedule.

A 'Machine Load Planner' is another type of Gantt chart used in industry. It consists of a large board with horizontal slots where each slot represents a separate machine. As jobs are scheduled for these machines, a job card, which has a length proportional to the time to be taken, is slotted in. By rearranging the job cards it is possible to maximise the use of each machine.

EXAMPLE 12

A small machine shop is manufacturing a product assembled from three components A, B and C. The available machines are a horizontal milling machine (H), a vertical milling machine (V), a centre lathe (T), a universal grinder (G) and a pedestal drill (D). Table 8.5 shows the sequence of operations required to be made on each component and the number of hours, in brackets, required to complete that operation.

TABLE 8.5

Sequence	A	B	C
1	$H(2)$	$T(2)$	$T(2)$
2	$G(2)$	$H(6)$	$V(4)$
3	$V(4)$	$T(4)$	$H(5)$
4	$G(3)$	$G(4)$	$T(3)$
5	$D(2)$		
Total hours	13	16	14

Construct a Gantt chart which shows how individual machines could be loaded.

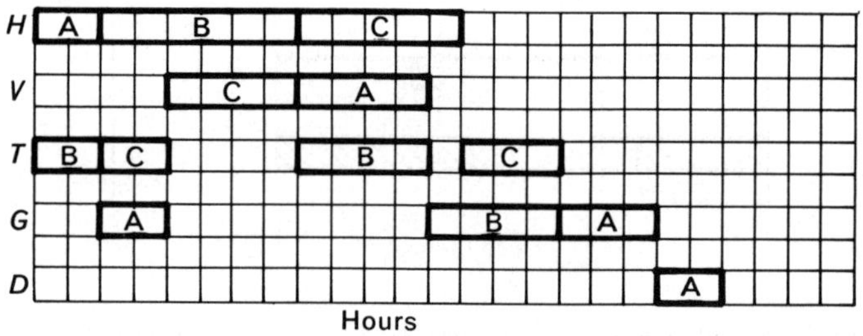

Fig. 8.14

Fig. 8.14 shows the sequence for each machine when the components are machined according to the total time required. That is B is completed, then C and finally A.

Fig. 8.15 shows one possible alternative showing a reduction of 3 hours to complete one product.

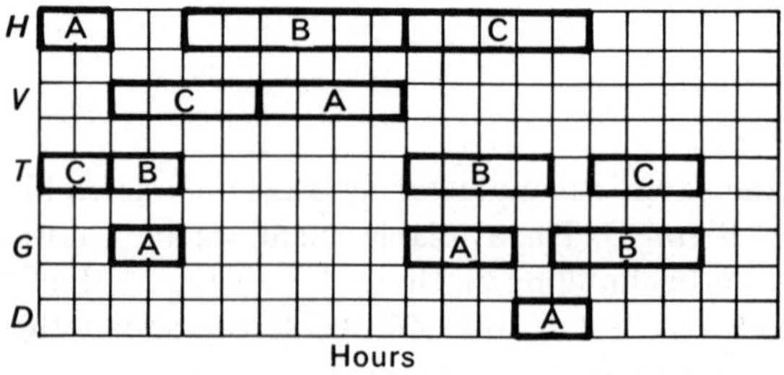

Fig. 8.15

THE COST OF OWNING MATERIAL

In order to make components a factory must not only pay for the required materials but also pay to have the materials stored. Hence we have

Total cost = Purchase price + purchasing costs + holding costs

where the **purchase price** is an amount assumed to be constant in any calculation and in the following example each unit has a constant price of £1.07.

The **purchasing costs** are required to place and service an order. In the following example the cost is £15 per order. Thus one large order will cost £15 to place, whereas the cost of three separate orders for the same goods will total £45.

As stock is used throughout the year the **holding costs** will fall. The cost is calculated on the average stock held over a period of time and in the following example it is taken as 20% of the purchase price.

EXAMPLE 13

A factory consumes 105 units of stock every 4 weeks; that is 1260 units for a 48 week year. The unit cost is £1.07, the ordering cost is £15 per order and the cost of holding or storing one unit is £0.214.

Make a list of the total costs involved when ordering the units in one lot, two lots, three lots etc., and hence determine the most economical method.

Since the purchase price of £1.07 per unit is constant it is unnecessary to consider it in the following table of variable costs.

Quantity purchased per order	No. of orders	Average quantity held over 1 year	Annual purchasing cost	Annual holding cost	Total cost
1260	1	630	£15	£135	£150
630	2	315	£30	£67	£97
420	3	210	£45	£45	£90←
315	4	158	£60	£34	£94
252	5	126	£75	£27	£102
210	6	105	£90	£22	£112
180	7	90	£105	£19	£124

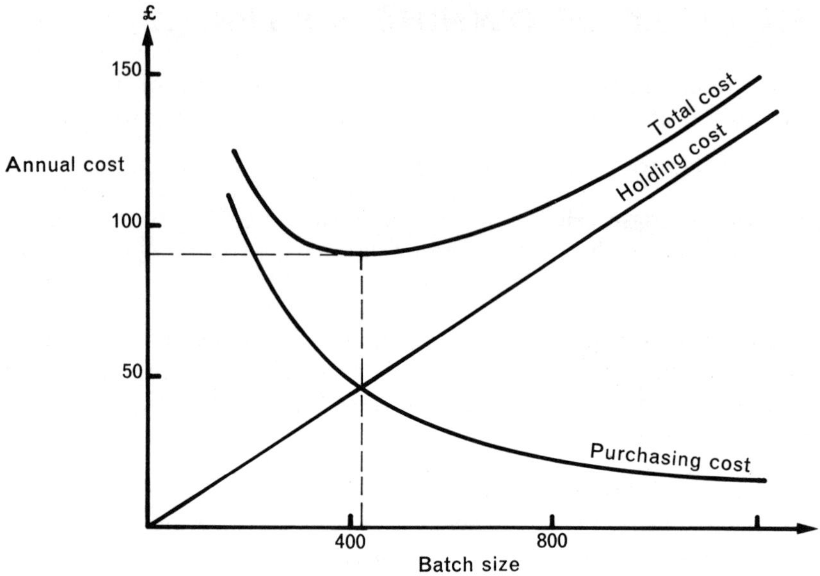

Fig. 8.16

From the tabulated figures, or from the graph of them shown in Fig. 8.16, it can be seen that the total cost, excluding the purchase price of $1260 \times £1.07$, is nearest a minimum when three batches, each of 420 units, are purchased throughout the year. If we let

Batch size $= B$

Purchasing costs per order $= P$

Holding costs per unit of time $= H$

Rate of usage of units in unit time $= R$

we have: Average stock held $= \frac{1}{2}B$

Annual purchasing costs $= \dfrac{P}{B} \times R$

Annual holding costs $= \frac{1}{2}B \times H$

$\therefore$ Acquisition cost, $A = \dfrac{PR}{B} + \frac{1}{2}BH$

For a minimum cost, $\dfrac{dA}{dB} = 0$

i.e. $$-\frac{PR}{B^2} + \tfrac{1}{2}H = 0$$

.Hence. for a minimum we require $\boxed{B = \sqrt{\dfrac{2PR}{H}}}$

Applying this formula to Example 13 we have:

$$B = \sqrt{\frac{2 \times 15 \times 1260}{20\% \times 1.07}} = 420 \text{ units}$$

This confirms the previous answer of ordering three batches, each of 420 units.

NOMOGRAMS

The construction of nomograms to evaluate formulae of all kinds is used in many engineering and manufacturing problems.

The simplest nomograms are based on the theorem that if three parallel lines AB, CD and EF are cut by a transversal PQR, see Fig. 8.17, then

$$\text{CQ} = \frac{k.\text{AP} + h.\text{ER}}{k + h}$$

Given a nomogram of this type we place a ruler at two chosen values of P and R and obtain the required corresponding result at Q.

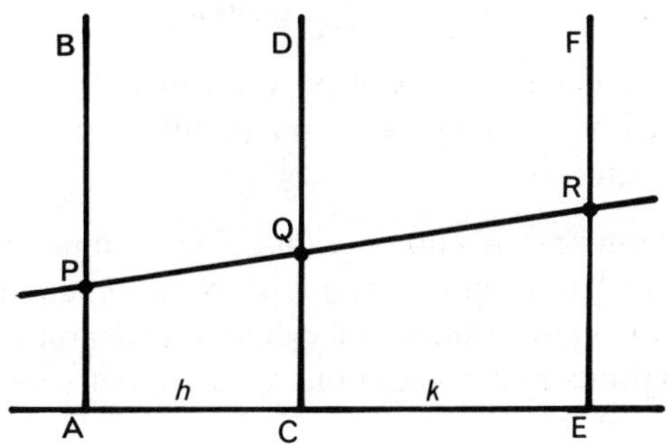

Fig. 8.17

As an illustration Fig. 8.18 shows a nomogram which evaluates $z = 2x+y$. In Fig. 8.18(a) the three parallel lines are equidistant, that is $h = k$. The x-line is scaled such that 2 units represent the number 10 while 1 unit represents the number 10 on the y-line. Since $2(10)+20 = 40$ we can fix 40 on the z-line by joining the two top points 10 and 20 with a ruler. The z-line can now be subdivided.

Fig. 8.18(b) shows an alternative method where the distances between the parallel lines are now in the ratio of 1 to 2; that is $h:k = 1:2$. This method gives greater accuracy since the scales on both the x and y-lines become the same resulting in twice the distance being allocated to each number on the y-line. Fewer numbers per unit length are also required on the z-line.

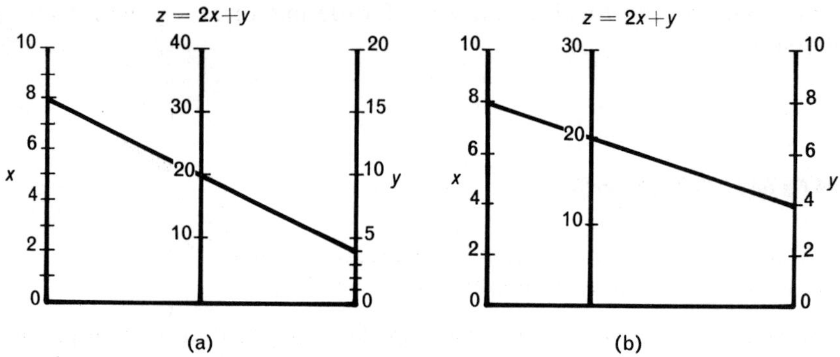

Fig. 8.18

Logarithmic scales can be constructed on the parallel lines. For example if we wish to construct a nomogram which connects $z = xy$ we rewrite this as

$$\log z = \log x + \log y$$

The logarithmic scales are marked on the outer lines, either by using a slide rule or logarithmic paper, while the middle scale is graduated at half the scale of the outer lines.

It is possible to combine two nomograms. For example, four parallel lines appropriately scaled and representing r, V, S and h, could be used to find both the surface area and volume of a cylinder, given the radius and height. Grid type nomograms and nomograms with curved scales are also widely used.

Exercise 27

1) A company is starting a process for the mass manufacture of compression springs. The specifications require the free lengths of the springs to have $\mu = 40.000$ mm and $\sigma = 0.010$ mm.
With samples of size 5, use the specifications to calculate a central line and control limits for an $\bar{x}$-chart.

2) Small rollers are manufactured to a given diameter. Ten samples, each of size 5, are measured and the results noted.
Calculate a central line and control limits for an $\bar{x}$-chart and an $\bar{R}$-chart. Draw the charts and plot the first ten points on each. (Use Tables 8.2 and 8.4.)

Sample	1	2	3	4	5	6	7	8	9	10
	4.88	4.94	4.98	4.92	4.91	5.01	4.97	4.95	5.04	5.00
	4.89	4.94	5.15	5.09	5.09	5.01	4.92	4.99	5.05	4.93
	5.05	4.99	4.87	5.14	4.99	5.05	4.96	4.98	5.13	5.06
	4.94	4.89	5.05	4.91	5.09	4.82	5.03	4.97	4.93	4.93
	5.00	5.05	5.11	5.01	4.98	5.20	4.92	4.87	5.03	5.02

3) Components from an automatic lathe are being measured in samples of size 4. The table shows a summary of the average and range for each of ten samples. Calculate the control limits and construct the control charts for average and range.

Sample No.	1	2	3	4	5	6	7	8	9	10
Average ($\bar{x}$ mm)	33.09	33.09	33.10	33.07	33.12	33.08	33.06	33.08	33.08	33.06
Range (R mm)	0.06	0.02	0.09	0.10	0.04	0.03	0.06	0.05	0.05	0.04

4) The work done by a firm of installation engineers is very seasonal. The planned installations and the actual installations for the previous six months are shown below. Draw a Gantt progress chart as it would have appeared at the end of April.

	Jan.	Feb.	Mar.	Apr.	May	Jun.
Planned	1000	1500	2000	2000	3000	3500
Actual	800	1600	1800	2100	3000	3200

5) A factory is engaged in the batch production of 8000 units per annum. The purchasing costs are £12.50, the cost of producing one unit is £0.875 and the cost of storage is £0.20 per unit.
Calculate the optimum batch size and the total cost of this batch.

6) Because of the nature of the material used in its manufacture the cost of storing a component is 50% of its value. If the purchasing costs are £2 per order and the cost of each component is £1, calculate the most economic batch size for an annual output of 5000 units.

PROJECTS

1) Control charts for statistics other than the mean and range are widely used. Investigate and write a report on the type or types used by your employers or obtain information about other types from a suitable book in the library.

2) Construct a simple nomogram which gives satisfactory results to a problem from an engineering situation.

3) Collect and write notes on the various nomograms and/or Gantt charts used by a particular industry.

4) Investigate the difficulties and methods of random sampling in a production environment (foreman's fingers, etc.).

9. INTEGRATION

After reaching the end of this chapter you should be able to:-

1. *Define indefinite integration as the reverse process of differentiation.*
2. *Determine indefinite integrals of simple functions (see 5).*
3. *Recognise the need to include an arbitrary constant of integration.*
4. *Define* $\int_a^b y\,dx$ *as the area under the curve between the ordinates* $x = a$ *and* $x = b$.
5. *Determine the areas between the x-axis, the curve and given ordinates for functions similar to those in Chapter 1, and polynomials of degree three or less.*
6. *Sketch a typical incremental volume under the graph of a function listed in 5.*
7. *Determine the volume of the increment in 6 in a suitable notation, e.g.* $\pi y^2 dx$.
8. *State the limits of the variable defining the required volume.*
9. *Determine the required volume by summing the incremental volumes using definite integration.*

INTEGRATION AS THE INVERSE OF DIFFERENTIATION

We have previously discovered how to obtain the differential coefficients of various functions. Our objective in this section is to find out how to reverse the process. That is, being given the differential coefficient of a function we try to discover the original function.

If $$y = \frac{x^4}{4}$$

then $$\frac{dy}{dx} = x^3$$

or we may write $$dy = x^3\,dx$$

The expression $x^3\,dx$ is called the differential of $\dfrac{x^4}{4}$.

Reversing the process of differentiation is called integration.

It is indicated by using the integration sign $\int$ in front of the differential.

Thus, if: $$dy = x^3\,dx$$

then $$y = \int x^3\,dx = \frac{x^4}{4}$$

Similarly if $$y = \frac{x^5}{5}$$

then $\quad \dfrac{dy}{dx} = x^4$

or $\quad dy = x^4\,dx$

and reversing the process:

$$y = \int x^4\,dx = \frac{x^5}{5}$$

Also, if $\quad y = \dfrac{x^{n+1}}{n+1}$

then $\quad \dfrac{dy}{dx} = x^n$

or $\quad dy = x^n\,dx$

from which $\quad y = \displaystyle\int x^n\,dx = \frac{x^{n+1}}{n+1}$

$\dfrac{x^{n+1}}{n+1}$ is called the integral of $x^n\,dx$.

This rule applies to all indices, positive, negative and fractional except for $\int x^{-1}\,dx$ which is a special case.

THE CONSTANT OF INTEGRATION

We know that the differential of $\dfrac{x^2}{2}$ is $x\,dx$. Therefore if we are asked to integrate $x\,dx$, $\dfrac{x^2}{2}$ is one answer but it is not the only possible answer because $\dfrac{x^2}{2}+2$, $\dfrac{x^2}{2}+5$, $\dfrac{x^2}{2}+19$ etc. are all expressions whose differential is $x\,dx$. The general expression for $\int x\,dx$ is therefore $\dfrac{x^2}{2}+c$, where c is a constant known as the constant of integration. Every time we integrate the constant of integration must be added in.

Thus:

$$\boxed{\int x^n\,dx = \frac{x^{n+1}}{n+1}+c}$$

EXAMPLE 1

(a) $\quad \displaystyle\int x^5\,dx = \frac{x^{5+1}}{5+1}+c = \frac{x^6}{6}+c$

(b) $$\int x\,dx = \frac{x^{1+1}}{1+1}+c = \frac{x^2}{2}+c$$

(c) $$\int \sqrt{x}\,dx = \int x^{1/2}\,dx = \frac{x^{3/2}}{3/2}+c = \frac{2x^{3/2}}{3}+c$$

(d) $$\int \frac{dx}{x^3} = \int x^{-3}\,dx = \frac{x^{-2}}{-2}+c = -\frac{2}{x^2}+c$$

A constant coefficient may be taken outside the integral sign.

Thus:

$$\int 3x^2\,dx = 3\int x^2\,dx = 3\frac{x^3}{3}+c = x^3+c$$

The integral of a sum is the sum of their separate integrals.

EXAMPLE 2

$$\int (x^2+x)\,dx$$

Integrating each term separately:

The integral of x^2 is $\frac{x^3}{3}$ and the integral of x is $\frac{x^2}{2}$

(a) $$\int (x^2+x)\,dx = \frac{x^3}{3}+\frac{x^2}{2}+c$$

(b) $$\int (5x^4+3x^2+2x-6)\,dx = x^5+x^3+x^2-6x+c$$

(c) $$\int (2x+5)^2\,dx = \int (4x^2+20x+25)\,dx$$

$$= \frac{4x^3}{3}+\frac{20x^2}{2}+25x+c$$

$$= \frac{4x^3}{3}+10x^2+25x+c$$

Exercise 28

Integrate with respect to x,

1) x^2

2) x^8

3) $\sqrt{x}$

4) $\frac{1}{x^2}$

5) $\frac{1}{x^4}$

6) $3x^4$

7) $\frac{1}{\sqrt{x}}$

8) $5x^8$

9) x^2+x+3

10) $2x^3-7x-4$

11) $x^2-5x+\frac{1}{\sqrt{x}}+\frac{2}{x^2}$

12) $\frac{8}{x^3}-\frac{2}{x^2}+\sqrt{x}$

13) $(x-2)(x-1)$

14) $(x+3)^2$

15) $(2x-7)^2$

Evaluating the Constant of Integration

The value of the constant of integration may be found provided a corresponding pair of values of x and y are known.

EXAMPLE 3

The gradient of the curve which passes through the point (2, 3) is given by x^2. Find the equation of the curve.

We are given $$\frac{dy}{dx} = x^2$$

$\therefore$ $$y = \int x^2\,dx = \frac{x^3}{3}+c$$

We are also given that when $x = 2,\ y = 3.$

Substituting these values in $$y = \frac{x^3}{3}+c$$

we have $$3 = \frac{2^3}{3}+c$$

or $$3 = 2\tfrac{2}{3}+c$$

$\therefore$ $$c = \frac{1}{3}$$

Hence the equation of the curve is:

$$y = \frac{x^3}{3}+\frac{1}{3}$$

or $$y = \frac{1}{3}(x^3+1)$$

Exercise 29

1) The gradient of the curve which passes through the point (2, 3) is given by x. Find the equation of the curve.

2) The gradient of the curve which passes through the point (3, 8) is given by (x^2+3). Find the value of y when $x = 5$.

3) It is known that for a certain curve $\frac{dy}{dx} = 3-2x$ and the curve cuts the x-axis where $x = 5$. Express y in terms of x. State the length of the intercept on the y-axis.

4) Find the equation of the curve which passes through the point (1, 4) and is such that $\frac{dy}{dx} = 2x^2+3x+2$.

5) If $\frac{dp}{dt} = (3-t)^2$ find p in terms of t given that $p = 3$ when $t = 2$.

6) A curve passes through the point (1, -2) and is such that $\frac{dy}{dx} = 3x^2-4x+1$. Prove that the equation of the curve may be written as $y = (x^2+1)(x-2)$.

7) The gradient of a curve is $ax+b$ at all points, where a and b are constants. Find the equation of the curve given that it passes through the points (0, 4) and (1, 3) and that the tangent at (1, 3) is parallel to the x-axis.

8) A curve passes through the point (0, 1) and is such that at every point of the curve $\frac{dy}{dx} = x^2$. Sketch the curve.

9) A curve passes through the point (2, 3) and is such that $\frac{dy}{dx} = 2-x^2$.

Find the value of y when $x = 1$.

THE DEFINITE INTEGRAL

It has been shown that $\int x^n \, dx = \frac{x^{n+1}}{n+1}+c$.

Since the expression contains an arbitrary constant c, the value of which is not known, it is called an indefinite integral.

A definite integral has a specific numerical answer without an unknown constant. An example of a definite integral is $\int_a^b x^n \, dx$. a and b are called limits, a being the lower limit and b the upper limit. The method of evaluating a definite integral is shown in the following examples.

EXAMPLE 4

Find the value of $\int_2^3 x^2 \, dx$.

$$\int_2^3 x^2 \, dx = \left[\frac{x^3}{3}+c\right]_2^3$$

$$= (\text{value of } \frac{x^3}{3}+c \text{ when } x \text{ is put equal to 3})$$

$$-(\text{value of } \frac{x^3}{3}+c \text{ when } x \text{ is put equal to 2})$$

$$= \left(\frac{3^3}{3}+c\right) - \left(\frac{2^3}{3}+c\right)$$

$$= \frac{27}{3}+c-\frac{8}{3}-c$$

$$= \frac{19}{3} = 6.33$$

In integration the use of the square brackets, as in the above solution, has a specific meaning, that is *the integration of each term has been completed and the next step is to substutite the values of the limits for* x.

We should also note that the constant c cancelled out. This will always happen and in solving definite integrals it is usual to omit c as shown in the next example.

EXAMPLE 5

Find the value of $\int_1^2 (3x^2-2x+5) \, dx$

$$\int_1^2 (3x^2-2x+5) = \left[x^3-x^2+5x\right]_1^2$$

$$= (2^3-2^2+5\times 2)-(1^3-1^2+5\times 1)$$

$$= 14-5 = 9$$

Exercise 30

Evaluate the following definite integrals:

1) $\int_1^2 x^2\,dx$

2) $\int_2^3 (2x+3)\,dx$

3) $\int_0^2 (x^2+3)\,dx$

4) $\int_1^2 (3x^2-4x+3)\,dx$

5) $\int_1^2 x(2x-1)\,dx$

6) $\int_0^2 \sqrt{x}\,dx$

7) $\int_1^3 \frac{1}{x^2}\,dx$

8) $\int_2^4 (x-1)(x-3)\,dx$

AREA UNDER A CURVE

Suppose that we wish to find the shaded area shown in Fig. 9.1. P, whose co-ordinates are (x, y) is a point on the curve.

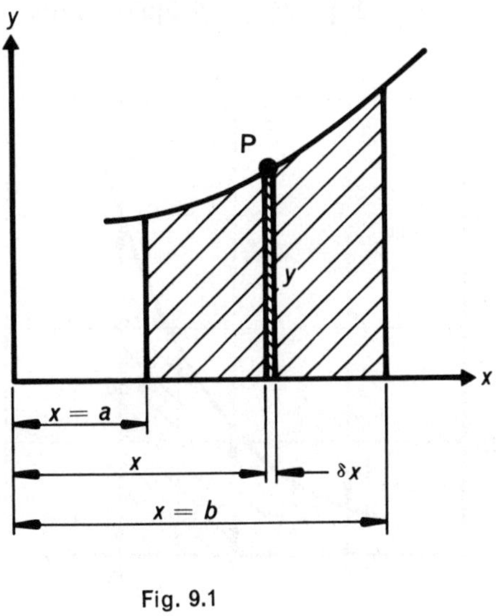

Fig. 9.1

Let us now draw, below P, a vertical strip whose width δx is very small. Since the width of the strip is very small we may consider the strip to be a rectangle with height y. Hence the area of the strip is approximately $y \times \delta x$. Such a strip is called an elementary strip and we will consider that the shaded area is made up from many elementary strips. Hence the required area is the sum of all the elementary strip areas between the values $x = a$ and $x = b$. In mathematical notation this may be stated as:

$$\text{Area} = \sum_{x=a}^{x=b} y \times \delta x \qquad \text{approximately.}$$

The process of integration may be considered to sum up an infinite number of elementary strips and hence gives an exact result.

$$\therefore \qquad \text{Area} = \int_a^b y\,\mathrm{d}x \qquad \text{exactly.}$$

EXAMPLE 6

Find the area bounded by the curve $y = x^3+3$, the x-axis and the lines $x = 1$ and $x = 3$.

$$\text{Area} = \int_1^3 (x^3+3)\,\mathrm{d}x = \left[\frac{x^4}{4}+3x\right]_1^3$$

$$= \left[\frac{3^4}{4}+3\times 3\right] - \left[\frac{1^4}{4}+3\times 1\right]$$

$$= 29\tfrac{1}{4} - 3\tfrac{1}{4} = 26 \text{ square units.}$$

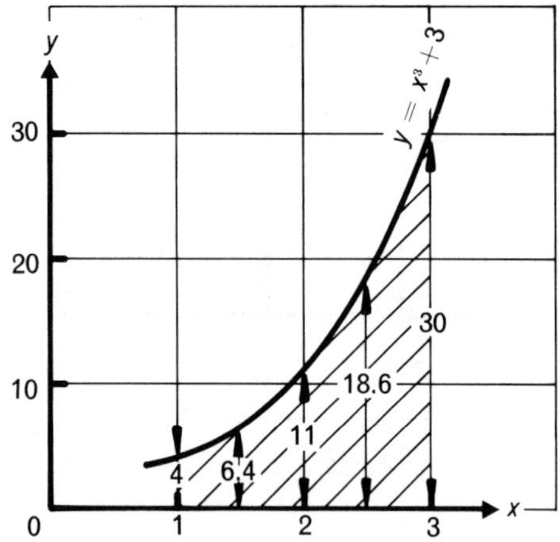

Fig. 9.2

The area we have obtained is shown shaded in Fig. 9.2. We will check the result by using the trapezoidal rule. Five ordinates have been taken and their lengths are as shown: we must not expect great accuracy from the use of only five ordinates but the result will serve as a check on the result obtained by integration.

The trapezoidal rule is:

$$\text{Area} = \text{width of strips} \times \left\{\frac{1}{2}(\text{sum of the first and last ordinates}) + (\text{sum of the remaining ordinates})\right\}$$

$$= 0.5 \times \left\{\tfrac{1}{2}(4+30)+(6.4+11+18.6)\right\}$$

$$= 26.5 \text{ square units.}$$

NEGATIVE AREAS

EXAMPLE 7

Sketch the curve $y = (x-1)(x+2)$ and hence find the area under the curve $y = (x-1)(x+2)$ between $x = -3$ and $x = 2$ and interpret the result.

The curve is sketched in Fig. 9.3 and it will be seen that it cuts the y-axis at $x = -2$ and $x = 1$. The area consists of three distinct parts as shown in the diagram, and each area will be calculated separately.

$$\text{Area } P = \int_{-3}^{-2} (x^2+x-2)\,dx = \left[\frac{x^3}{3}+\frac{x^2}{2}-2x\right]_{-3}^{-2}$$

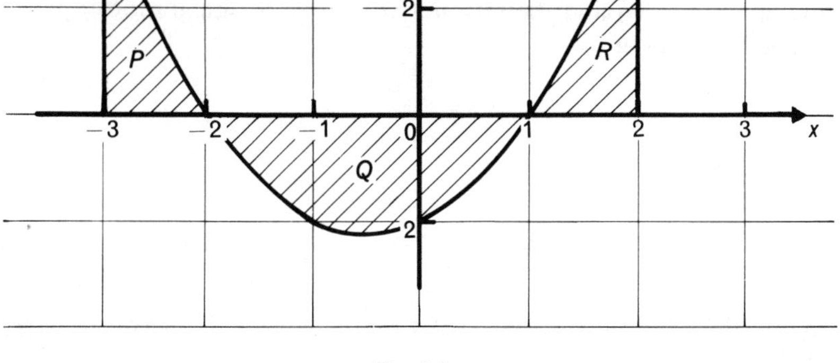

Fig. 9.3

$$= \left(\frac{(-2)^3}{3}+\frac{(-2)^2}{2}-2(-2)\right)-\left(\frac{(-3)^3}{3}+\frac{(-3)^2}{2}-2(-3)\right)$$

$$= 1.833 \text{ square units}$$

$$\text{Area } Q = \int_{-2}^{1} (x^2+x-2)\,dx = \left[\frac{x^3}{3}+\frac{x^2}{2}-2x\right]_{-2}^{1}$$

$$= \left(\frac{(1)^3}{3}+\frac{(1)^2}{2}-2(1)\right)-\left(\frac{(-2)^3}{3}+\frac{(-2)^2}{2}-2(-2)\right)$$

$$= -4.500 \text{ square units}$$

$$\text{Area } R = \int_{1}^{2} (x^2+x-2)\,dx = \left[\frac{x^3}{3}+\frac{x^2}{2}-2x\right]_{1}^{2}$$

$$= \left(\frac{(2)^3}{3}+\frac{(2)^2}{2}-2(2)\right)-\left(\frac{(1)^3}{3}+\frac{(1)^2}{2}-2(1)\right)$$

$$= 1.833 \text{ square units}$$

Note that the areas P and R which lie above the x-axis are positive areas whilst the area Q which lies below the x-axis is a negative area. If we integrate between the limits $x = -3$ and $x = 2$ we get:

$$\int_{-3}^{2} (x^2+x-2)\,dx = \left[\frac{x^3}{3}+\frac{x^2}{2}-2x\right]_{-3}^{2}$$

$$= \left(\frac{(2)^3}{3}+\frac{(2)^2}{2}-2(2)\right)-\left(\frac{(-3)^3}{3}+\frac{(-3)^2}{2}-2(-3)\right)$$

$$= -0.833 \text{ square units}$$

Now area P+area Q+area R $= 1.833-4.500+1.833$
$= -\ 0.833$ square units.

Hence when we integrate between the limits $x = -3$ and $x = 2$, the area given by the integration is the *net* area, that is, the signs of the various areas are taken into account.

Before attempting to find the area under a curve by integration it pays to first sketch the curve.

Exercise 31

1) Find the area under the curve $y = x^2+3$ between $x = 1$ and $x = 3$.

2) Find the area under the curve $y = x^2-2x+2$ between $x = 0$ and $x = 2$.

3) Find the area between the curve $y = (x-1)(x-2)$ and the x-axis.

4) Find the area bounded by the curve $y = (1+x)(2-x)$ and the x-axis.

5) Sketch the curve $y = x(x-1)(x-2)$ and find the two areas between the curve and the x-axis.

6) Show that $\int_{-2}^{2} x(x^2-4)\,dx = 0$, and interpret the result graphically.

7) Find the area enclosed between the curve $y^2 = 4x$ and the line $y = x$.

8) Find the area between the curve $y = \frac{1}{x}+x$, the x-axis and the lines $x = 1$ and $x = 2$.

9) Show that $\int_{0}^{2\pi} \sin x\,dx = 0$ and explain the result with reference to a graph.

10) Find the area bounded by the curve $y = \cos x$, the x-axis, and the lines $x = 0$ and $x = \frac{\pi}{3}$.

11) Calculate the area enclosed between the x-axis and that part of the curve $y = 10+3x-x^2$ which lies above it.

12) Show that the curve $y = 4x-\frac{32}{x^2}$ crosses the x-axis at the point where $x = 2$. Calculate the area between the curve, the x-axis and the ordinate $x = 4$.

VOLUMES OF REVOLUTION BY INTEGRATION

If the area under the curve APB (Fig. 9.4) is rotated one complete revolution about the x-axis, then the volume swept out is called a volume of revolution.

The point P, whose co-ordinates are (x, y) is a point on the curve AB.

Let us consider, below P, a thin slice whose width is δx. Since the width of the slice is very small we may consider the slice to be a cylinder of radius y. The volume of this slice is approximately $\pi y^2.\delta x$. Such a slice is called an elementary slice and we will consider that the volume of revolution is made up from many such elementary slices. Hence the complete volume of revolution is the sum of all the elementary slices between the values of $x = a$ and $x = b$.

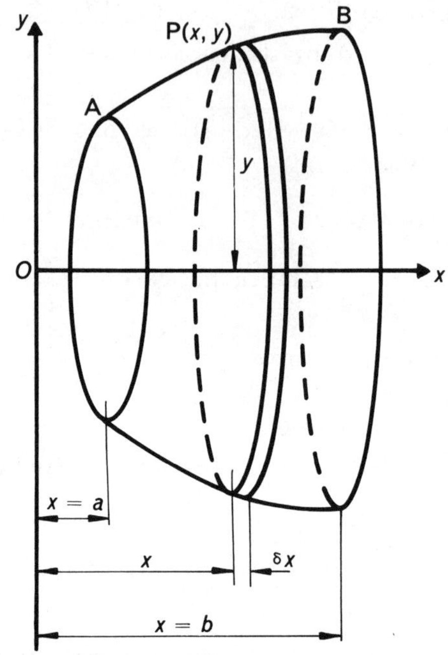

Fig. 9.4

In mathematical notation this may be stated as

$$\sum_{x=b}^{x=a} \pi y^2 . \delta x \quad \text{approximately.}$$

As for areas, the process of integration may be considered to sum an infinite number of elementary slices and hence it gives an exact result.

$$\therefore \qquad \text{Volume of revolution} = \int_a^b \pi y^2 \, dx \quad \text{exactly}$$

EXAMPLE 8

The area between the curve $y = x^2$, the x-axis and the ordinates $x = 1$ and $x = 3$ is rotated about the x-axis. Find the volume of revolution.

As when finding areas it is recommended that a sketch is made of the required volume. Fig. 9.5 shows a sketch of the required volume.

$$\text{Required volume of revolution} = \int_1^3 \pi y^2 \, dx$$

$$= \int_1^3 \pi (x^2)^2 \, dx$$

$$= \pi \int_1^3 x^4 \, dx$$

$$= \pi \left[\frac{x^5}{5} \right]_1^3$$

$$= \pi \left\{ \frac{3^5}{5} - \frac{1^5}{5} \right\}$$

$$= 152.1 \text{ cubic units.}$$

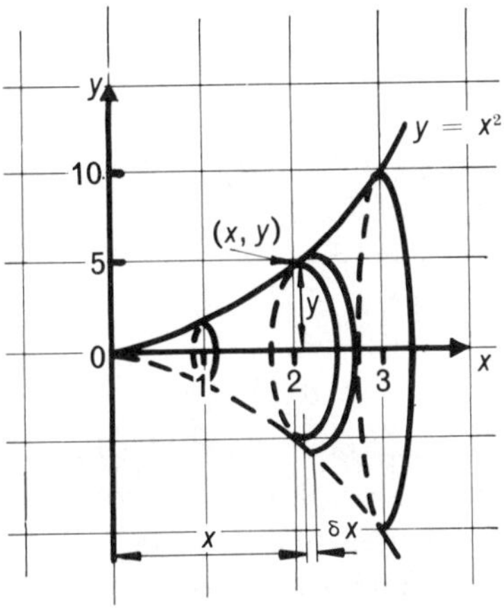

Fig. 9.5

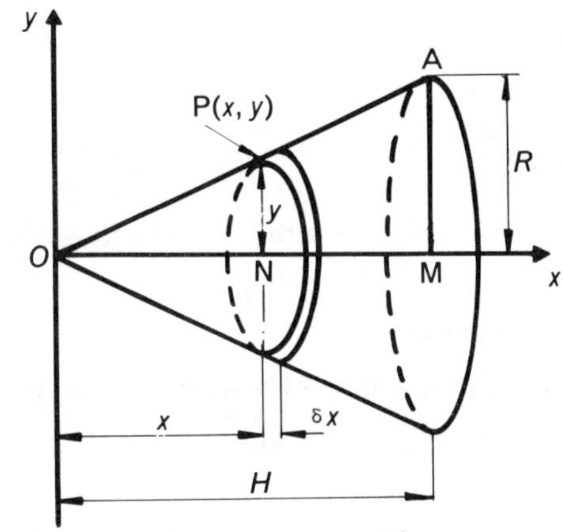

Fig. 9.6

EXAMPLE 9

Find, by the calculus, the volume of a cone of base radius R and height H.

The first step is to set up the cone on suitable axes. For convienence the cone has been put with its polar axis lying along the x-axis as shown in Fig. 9.6.

Then the volume of the cone is the volume of revolution generated when the area OAM is rotated about the x-axis.

We need an equation connecting x and y. This may be found by considering the similar triangles OPN and OAM.

Then $$\frac{\text{PN}}{O\text{N}} = \frac{\text{AM}}{O\text{M}}$$

or $$\frac{y}{x} = \frac{R}{H}$$

or $$y = \frac{R}{H} \cdot x$$

$$\text{Now the required volume} = \int_0^H \pi y^2\,dx$$

$$= \int_0^H \pi\left(\frac{R}{H}.x\right)^2 dx$$

$$= \pi\frac{R^2}{H^2}\int_0^H x^2\,dx$$

$$= \pi\frac{R^2}{H^2}\left[\frac{x^3}{3}\right]_0^H$$

$$= \pi\frac{R^2}{H^2}\left\{\frac{H^3}{3}-\frac{0^3}{3}\right\}$$

$$= \frac{1}{3}\pi R^2 H$$

which verifies a formula you know already.

EXAMPLE 10

Find the volume of a sphere of radius R.

The volume of a sphere may be considered to be a volume of revolution generated by the revolution of a semi-circular area about its boundary diameter. Suitable axes must now be chosen and these are shown in Fig. 9.7.

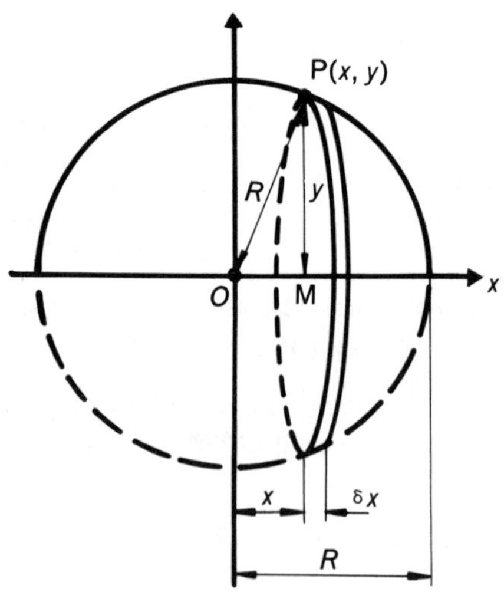

Fig. 9.7

The relationship connecting y and x may be found by considering the right-angled triangle OPM and applying the theorem of Pythagoras.

Hence $\qquad x^2+y^2 = R^2$

$\therefore \qquad y^2 = R^2-x^2$

Now the volume of revolution $= \int \pi y^2 \, \mathrm{d}x$

$$= \pi \int_{-R}^{R} (R^2-x^2) \, \mathrm{d}x$$

$$= \pi \left[R^2x - \frac{x^3}{3} \right]_{-R}^{R}$$

$$= \pi \left\{ \left(R^2(R) - \frac{R^3}{3} \right) - \left(R^2(-R) - \frac{(-R)^3}{3} \right) \right\}$$

$$= \frac{4}{3}\pi R^3$$

verifying a formula you should recognise.

Exercise 32

In Questions 1–5 find the volume generated about the x-axis of the given curves between the limits stated:

1) $y = x^3$ from $x = 0$ to $x = 2$.

2) $xy = 16$ from $x = 1$ to $x = 2$.

3) $x = 4\sqrt{y}$ from $x = 1$ to $x = 2$.

4) $y^3 = x^2$ from $x = 1$ to $x = 8$.

5) $y = \frac{x}{2}$ from $x = 0$ to $x = 4$.

6) Find the volume generated by revolving about the x-axis the portion of the curve $y = x(x-1)$ which lies below the x-axis.

7) The portion of the circle $x^2+y^2 = 25$ in the first quadrant is rotated about the x-axis. Find:

(a) the volume of a hemisphere of radius 50 mm,

(b) the volume of a section of the above hemi-sphere cut off by the planes distant 20 mm and 30 mm from the plane base.

8) A bucket has a radius of 100 mm at the base, and 200 mm at the top. It is 200 mm deep and the sides slope uniformly. Show that it may be considered as being formed by the revolution of the line $y = \frac{x}{2}+100$ about the x-axis from $x = 0$ to $x = 200$. Hence find the capacity of the bucket, when full, in litres.

10. CENTROIDS OF AREAS

On reaching the end of this chapter you should be able to:-

1. *Sketch a given area including a typical incremental area whose centroid is known.*
2. *Determine the moment of the increment in 1 about a specified axis in the plane of the area.*
3. *Sum the moments of the increments between given limits by definite integration.*
4. *Determine the given area.*
5. *Define centroid as the ratio of the total mom[illegible] determined in 3 to the total area determined ir*
6. *Calculate the distance of the centroid from [illegible] given axis.*
7. *Calculate the centroids of a rectangle, trian[illegible] circle and sector of a circle.*

CENTROIDS OF AREAS

We know that the moment of a force F about the line CD (Fig. 10.1) is given by $F\times d$.

Similarly we say that the first moment of the area A about the line CD (Fig. 10.2) is given by $A\times d$, the point G being the centroid of the area.

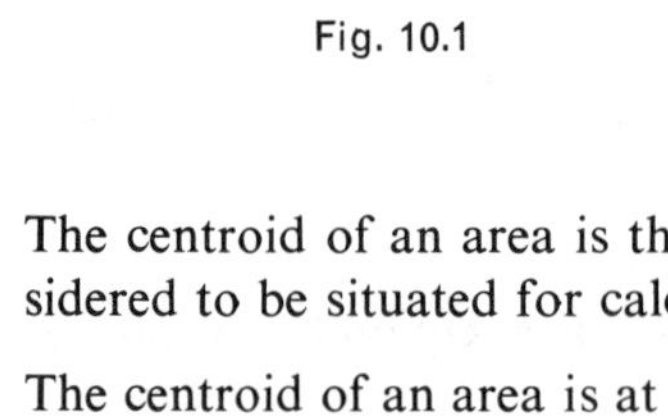

Fig. 10.1

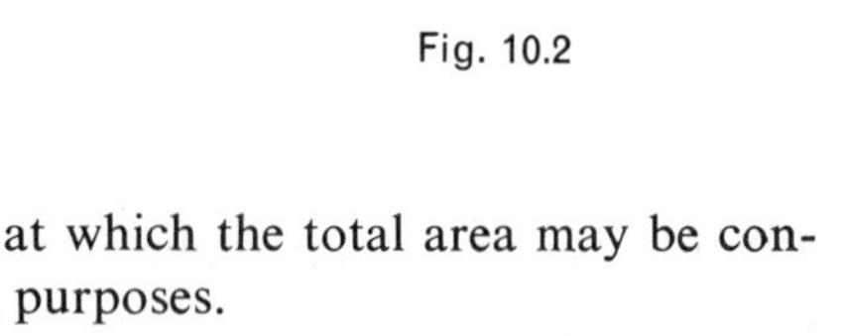

Fig. 10.2

The centroid of an area is the point at which the total area may be considered to be situated for calculation purposes.

The centroid of an area is at the point which corresponds to the centre of gravity of a lamina of the same shape as the area. A thin flat sheet of steel of uniform thickness is an example of a lamina.

We know the position of the centres of gravity of the commonest shapes of laminae: for example the centre of gravity of a thin flat circular disc is at the centre of the disc. Hence we may say that the centroid of a circular area is at the centre of the area.

It is often possible to deduce the position of the centroid by the symmetry of the area. For example, in the case of a rectangle the centroid will be at the intersection of the centre-lines, as shown in Fig. 10.3, as the area is symmetrical about either of the centre-lines.

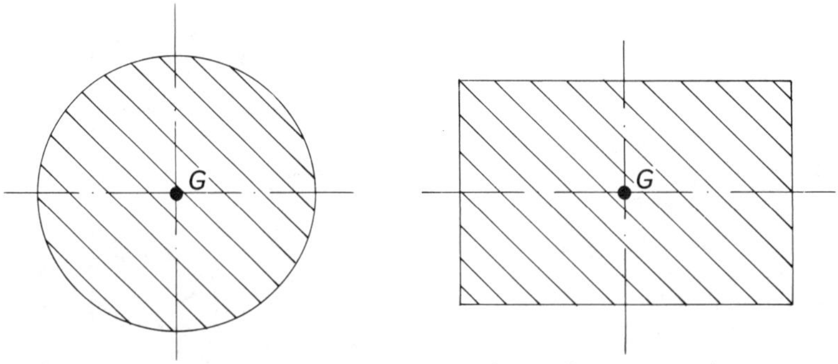

Positions of the centroid G for a circular and a rectangular area.

Fig. 10.3

It is often necessary to find the position of the centroid of a *composite area*. This is an area which is made up from common shapes. To determine the position of the centroid we use two reference axes Ox and Oy. The location of the centroid is then given by the co-ordinates $\bar{x}$ and $\bar{y}$ as shown in Fig. 10.4.

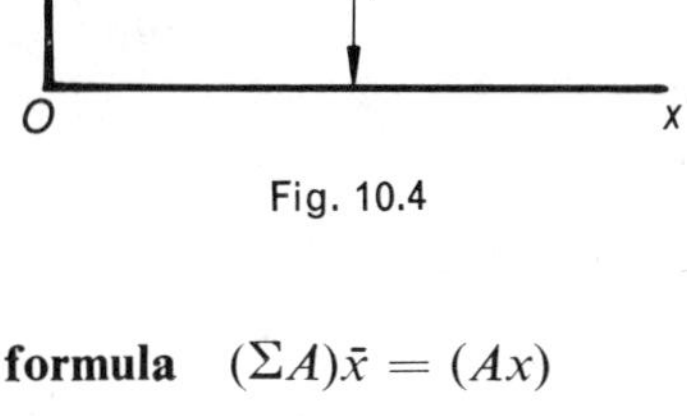

Fig. 10.4

To find $\bar{x}$ we use the formula $(\Sigma A)\bar{x} = (Ax)$

where ΣA is the sum of the component areas

and ΣAx is the sum of the first moments of the component areas about the axis Oy.

To find $\bar{y}$ we use the formula $(\Sigma A)\bar{y} = (\Sigma Ay)$

The following example will show you how to use these formulae.

EXAMPLE 1

Find the position of the centroid shown in Fig. 10.5.

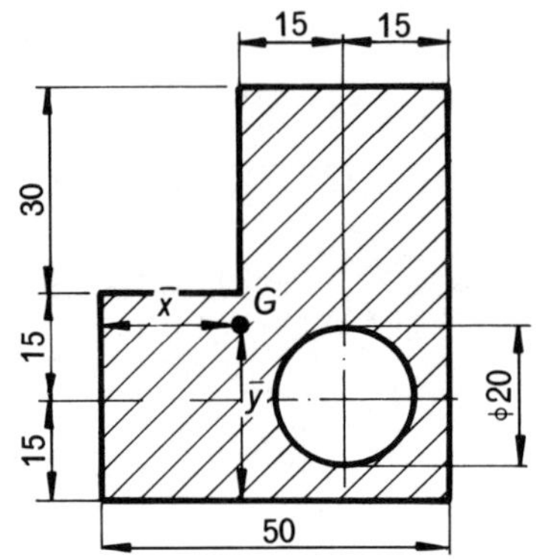

Fig. 10.5

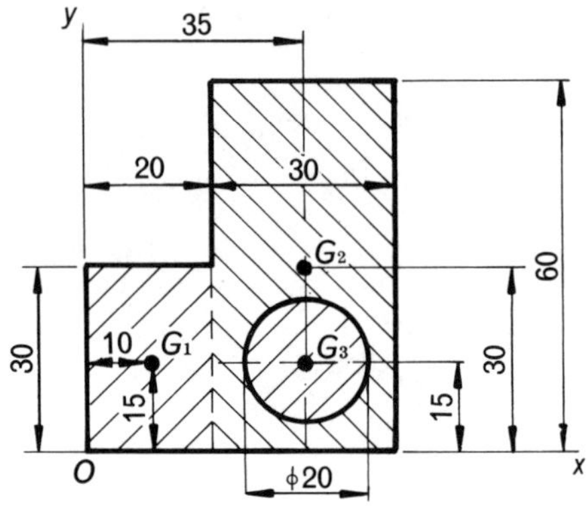

Fig. 10.6

For convenience the reference axes Ox and Oy have been chosen as shown in Fig. 10.6. It will also be seen that the given area has been divided into three component areas whose centroids are G_1, G_2 and G_3. The dimensions of each component area and the location of each centroid from the axes Ox and Oy are clearly shown. A diagram of this type is essential for this and similar problems — any attempt to obtain these dimensions mentally usually results in one or more errors. Remember the circle is a 'missing' area and must be taken as negative in the calculations.

It helps to simplify the arithmetic and hence reduce errors by showing the solution in tabular form.

Area	Distance to centroid		Moment of area	
	from Oy	from Ox	about Oy	about Ox
A	x	y	Ax	Ay
$20\times30 = 600$	10	15	$600\times10 = 6000$	$600\times15 = 9000$
$30\times60 = 1800$	35	30	$1800\times35 = 63\,000$	$1800\times30 = 54\,000$
$-\pi\times10^2 = -310$	35	15	$-310\times35 = -10\,900$	$-310\times35 = -4700$
$\Sigma A = 2090$			$\Sigma Ax = 58\,100$	$\Sigma Ay = 58\,300$

Hence $$\bar{x} = \frac{\Sigma Ax}{\Sigma A} = \frac{58\,100}{2090} = 27.8 \text{ mm}$$

and $$\bar{y} = \frac{\Sigma Ay}{\Sigma A} = \frac{58\,300}{2090} = 27.9 \text{ mm}$$

FINDING CENTROIDS USING INTEGRATION

There are many shapes which cannot be divided up exactly into either rectangles or circles. These more complicated shapes can be split up into elemental strips which approximate to rectangles. Then by finding the dimensions of each strip, etc. the formulae $\bar{x} = \frac{\Sigma Ax}{\Sigma A}$ and $\bar{y} = \frac{\Sigma Ay}{\Sigma A}$ may be used to find the approximate values of $\bar{x}$ and $\bar{y}$.

However if we know, or can find, the equation of y in terms of x, by setting up the area on suitable axes, the numerical summing up may be achieved by integration and exact results obtained.

The following examples will explain how integration is used to find a centroid.

EXAMPLE 2

Find, by integration, the value of $\bar{x}$ (i.e. the distance of the centroid from the left hand edge) for the rectangle shown in Fig. 10.7.

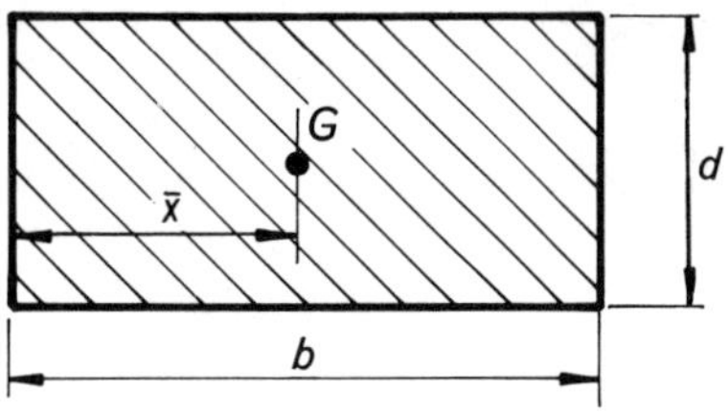

Fig. 10.7

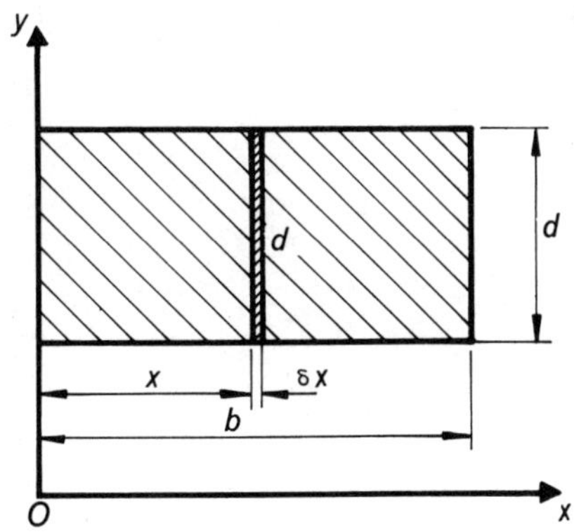

Fig. 10.8

The rectangle must be set up on suitable axes. Fig. 10.8 shows a sketch of the arrangement and includes the usual elementary strip of very small width δx and length d.

Now the area of the rectangle $= \Sigma A$

$$= \text{sum of the elementary strip areas}$$

$$= \sum_{x=0}^{x=b} d.\delta x$$

$$= \int_0^b d.\mathrm{d}x$$

$$= d\int_0^b 1.\mathrm{d}x$$

$$= d\Big[x\Big]_0^b = d(b-0) = b.d$$

Also the first moment of the rectangular area about the y-axis

$$= \Sigma\, A.x$$

= sum of the first moment of area of each of the elementary strips

= sum of: (area of strip × distance of its centroid from the y-axis)

$$= \sum_{x=0}^{x=b} (d.\delta)x$$

$$= \int_0^b d.x.\mathrm{d}x$$

$$= d\int_0^b x.\mathrm{d}x$$

$$= d\left[\frac{x^2}{2}\right]_0^b$$

$$= d\left(\frac{b^2}{2}-\frac{0^2}{2}\right)$$

$$= \frac{b^2.d}{2}$$

Hence $$\bar{x} = \frac{\Sigma\, A.x}{\Sigma\, A} = \frac{b^2 d/2}{bd} = \frac{b}{2}$$

This is the result we would expect from the symmetry of the figure. The procedure followed in this example is always used when finding positions of centroids by integration.

EXAMPLE 3

Find the position of the centroid of a triangle of height H and base length B.

The triangle must be set up on suitable axes. Only experience enables a good choice. In this case it is convenient to turn the triangle through 90° so that the apex lies on the vertical y-axis and the base is parallel to it. In this instance the position of the x-axis is unimportant.

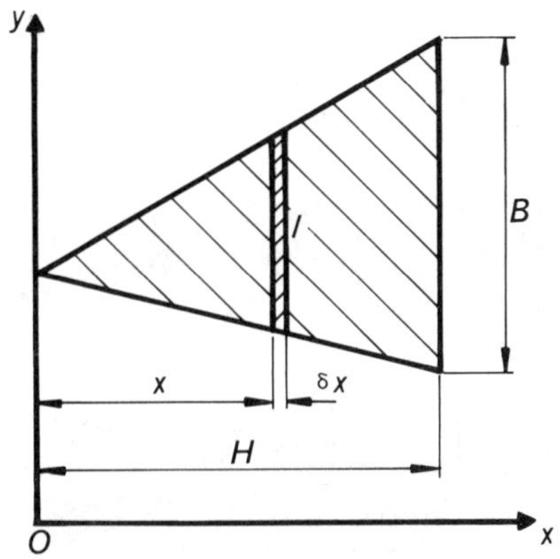

Fig. 10.9

Fig. 10.9 shows a sketch of the arrangement and includes the usual elementary strip area of very small width δx and length l. We shall need a connection between l and x and this may be found using similar triangles

giving $\dfrac{l}{x} = \dfrac{B}{H}$ from which $l = \dfrac{B}{H}.x.$

Now the area of the triangle $= \Sigma A$

$$= \text{sum of the elementary strip areas}$$

$$= \sum_{x=0}^{x=H} l.\delta x$$

$$= \int_0^H l.\mathrm{d}x$$

$$= \int_0^H \frac{B}{H}.x.\mathrm{d}x$$

$$= \frac{B}{H}\int_0^H x.\mathrm{d}x$$

$$= \frac{B}{H}\left[\frac{x^2}{2}\right]_0^H = \frac{B}{H}\left\{\frac{H^2}{2} - \frac{0^2}{2}\right\} = \frac{1}{2}BH$$

Also the first moment of the triangular area about the y-axis:

$$\Sigma\, Ax = \text{sum of the first moment of area of each of the elementary strips}$$

$$= \text{sum of (area of strip} \times \text{distance of its centroid from the } y\text{-axis)}$$

$$= \sum_{x=0}^{x=H} l.\delta x.x$$

$$= \int_0^H l.\mathrm{d}x.x$$

$$= \int_0^H \frac{B}{H}.x.\mathrm{d}x.x$$

$$= \frac{B}{H}\int_0^H x^2\,\mathrm{d}x$$

$$= \frac{B}{H}\left[\frac{x^3}{3}\right]_0^H = \frac{B}{H}\left\{\frac{H^3}{3}-\frac{0^3}{3}\right\} = \frac{1}{3}BH^2$$

Hence $\bar{x}$ $$= \frac{\Sigma\, Ax}{\Sigma\, A} = \frac{\frac{1}{3}BH^2}{\frac{1}{2}BH} = \frac{2}{3}H$$

It should be noted that this is independent of the base length B.

You may remember that the position of the centroid of a triangular area is as given in Fig. 10.10. The above calculations verify this.

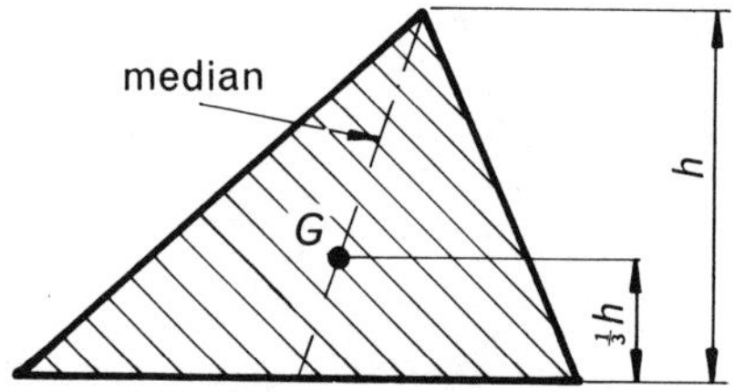

Fig. 10.10

EXAMPLE 4

Find the position of the centroid of the sector of a circle of radius R and angle 2α.

The sector area is shown in Fig. 10.11 located on suitable axes Ox and Oy. By symmetry the centroid G of the sector area lies on the centre-line OM, and its position along the line may be defined if we find the distance $\bar{x}$.

It is not convenient to divide the area into vertical (or even horizontal) elementary strip areas. Instead we consider elementary sector areas of small angle $\delta\theta$ (Fig. 10.12). Each of these elementary areas may be considered to approximate to a triangle having base length $R\times\delta\theta$ (since

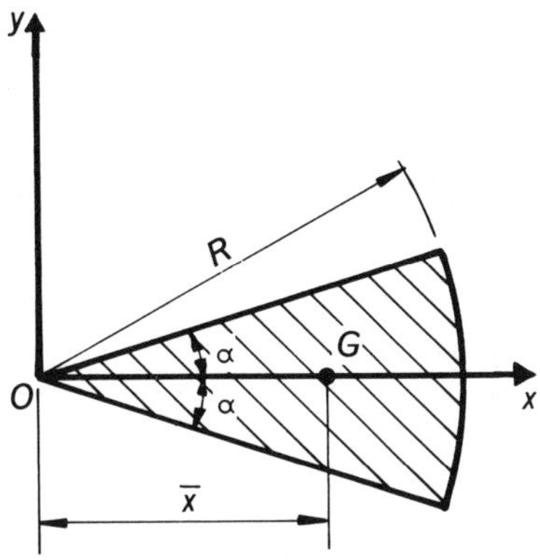

Fig. 10.11

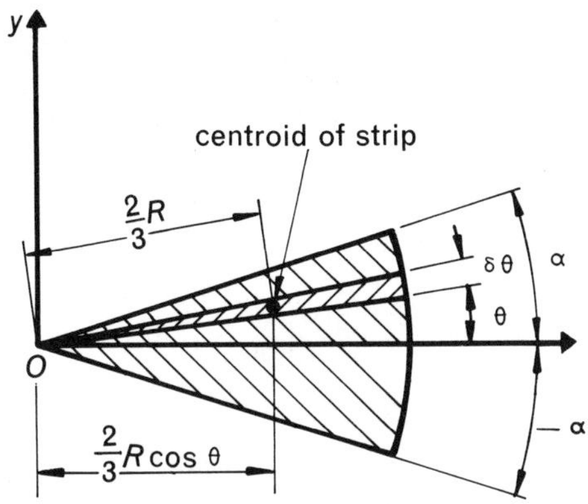

Fig. 10.12

length of arc = angle × radius, where the angle is in radians), and height R.

We must first find the area of the sector by summing the areas of the elementary sector areas.

Hence area of sector $= \Sigma A$

= sum of elementary sector areas

$$= \sum_{\theta=-\alpha}^{\theta=\alpha} \frac{1}{2}(R.\delta\theta)R \quad \text{since the area of a triangle is } \tfrac{1}{2}\text{(base)(height)}$$

$$= \int_{-\alpha}^{\alpha} \frac{1}{2}(R.\mathrm{d}\theta)R$$

$$= \tfrac{1}{2}R^2 \int_{-\alpha}^{\alpha} \mathrm{d}\theta$$

$$= \tfrac{1}{2}R^2\Big[\theta\Big]_{-\alpha}^{\alpha} = \frac{1}{2}R^2\{\alpha-(-\alpha)\}$$

$$= R^2\alpha$$

Now the first moment of the given sector area about the y-axis:

ΣAx = sum of the first moment of area of each of the elementary sector areas

= sum of (elementary area × distance of its centroid from the y-axis)

$$= \sum_{\theta=-\alpha}^{\theta=\alpha} (\tfrac{1}{2}R^2\delta\theta)(\tfrac{2}{3}R.\cos\theta)$$

$$= \int_{-\alpha}^{\alpha} \tfrac{1}{2}R^2.d\theta.\tfrac{2}{3}R.\cos\theta$$

$$= \tfrac{1}{3}R^3 \int_{-\alpha}^{\alpha} \cos\theta \, d\theta$$

$$= \tfrac{1}{3}R^3 \Big[\sin\theta\Big]_{-\alpha}^{\alpha}$$

$$= \tfrac{1}{3}R^3\{\sin\alpha - \sin(-\alpha)\}$$

$$= \tfrac{1}{3}R^3\{\sin\alpha + \sin\alpha\}$$

$$= \tfrac{2}{3}R^3 \sin\alpha.$$

Now

$$\bar{x} = \frac{\Sigma\, Ax}{\Sigma\, A}$$

$$= \frac{\tfrac{2}{3}R^3.\sin\alpha}{R^2\alpha}$$

$$= \frac{2}{3}R\frac{\sin\alpha}{\alpha}$$

It should be remembered that the angle α must be in radians.

EXAMPLE 5

Using the result obtained for the position of the centroid of the sector of a circle find:

(a) the position of the centroid of a quadrant of a circle of radius R.

(b) the position of the centroid of a semicircular area of radius R.

(a) If we set the quadrant area on suitable axes as shown in Fig. 10.13, then we may use the formula for the sector of a circle giving:

$$OG = \tfrac{2}{3}R\frac{\sin}{\alpha}\alpha \quad \text{where} \quad \alpha = \frac{\pi}{4} \text{ rad}$$

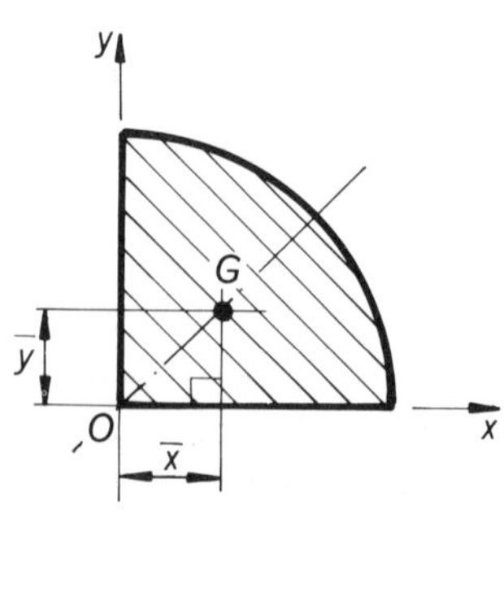

Fig. 10.13

From right-angled triangle OGM we have:

$$
\begin{aligned}
\bar{x} &= OM \\
&= OG\left(\cos\frac{\pi}{4}\right) \\
&= \tfrac{2}{3}R\,\frac{\sin\frac{\pi}{4}}{\frac{\pi}{4}}\left(\cos\frac{\pi}{4}\right) \\
&= \frac{2R(0.7071)(0.7071)4}{3\pi} \\
&= \frac{4R}{3\pi} \\
&= 0.424R
\end{aligned}
$$

Also by symmetry $\bar{y}$ will have the same value as $\bar{x}$.

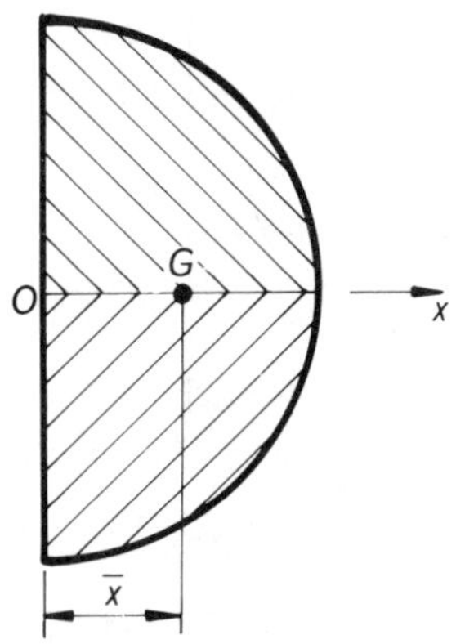

Fig. 10.14

(b) The semi-circular area shown in Fig. 10.14 can be considered to be two quadrants as indicated. It follows, therefore, that $\bar{x}$ for a semi-circular area is the same as that for a quadrant, i.e. $0.424R$.

By symmetry the centroid also lies on the horizontal centre-line Ox.

Exercise 33

In Questions 1–4 find the positions of the centroids of the cross-sectional areas shown, giving the distances in each case from the left hand edge (or extreme point) and above the bottom edge (or extreme point).

1)

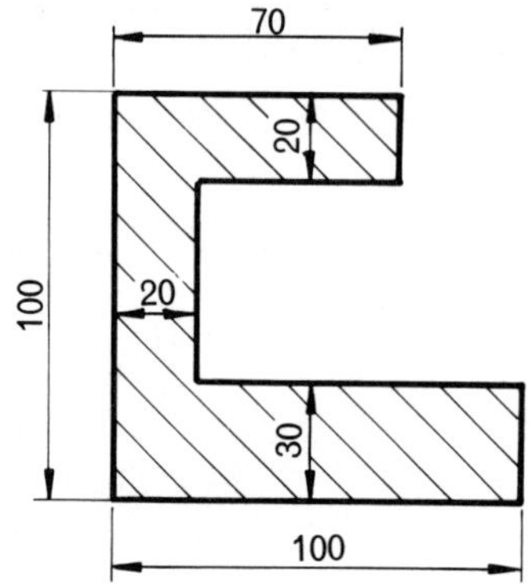

2)

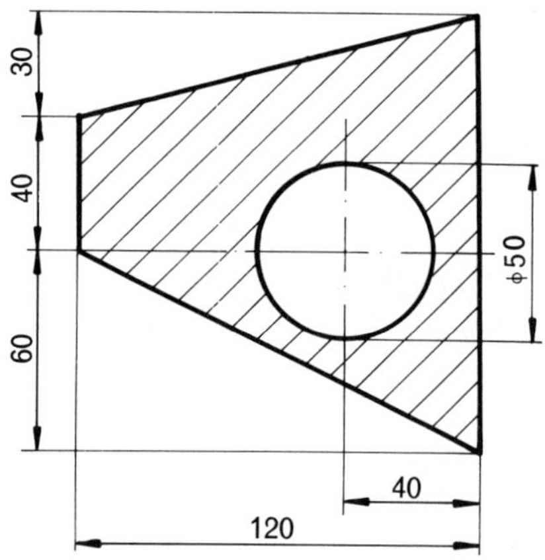

3)

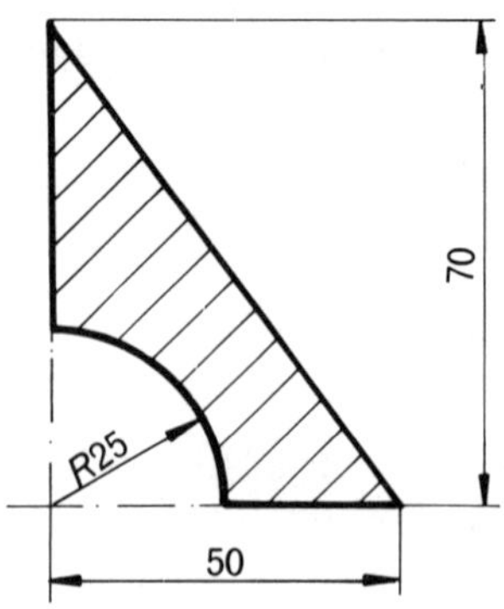

4)

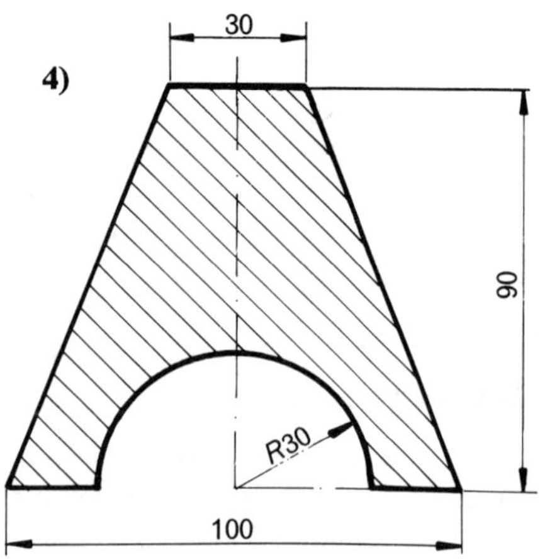

In Questions 5–8 find $\bar{x}$ and $\bar{y}$ as indicated on the diagrams.

5)

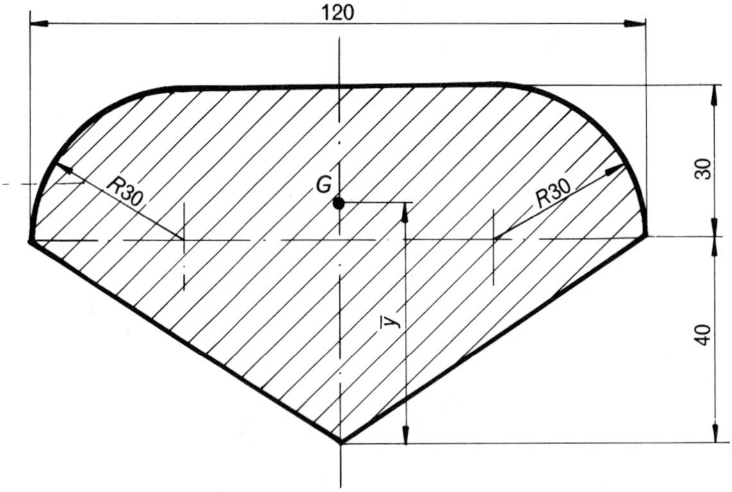

6)

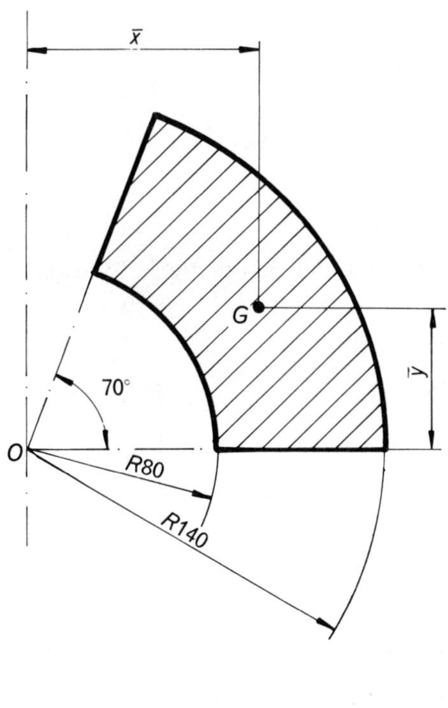

7)

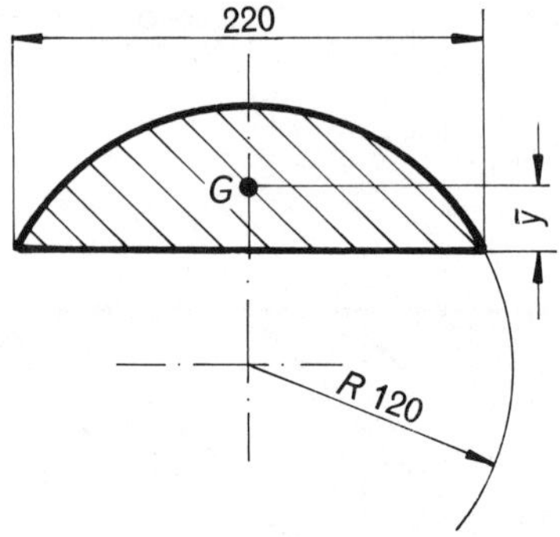

8)

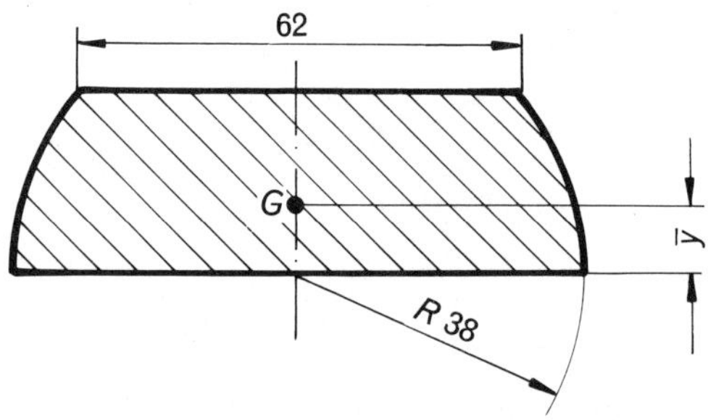

SUMMARY

Centroids of the more common shapes are as follows:

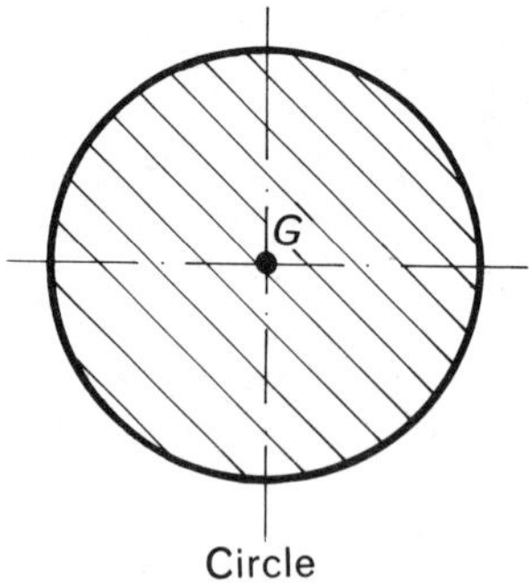

Circle

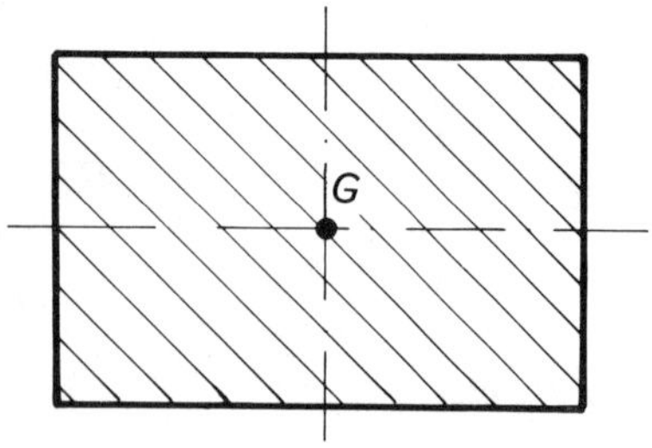

Rectangle

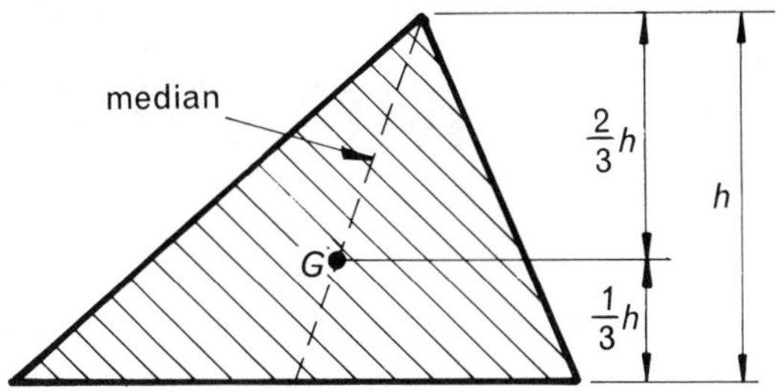

Triangle

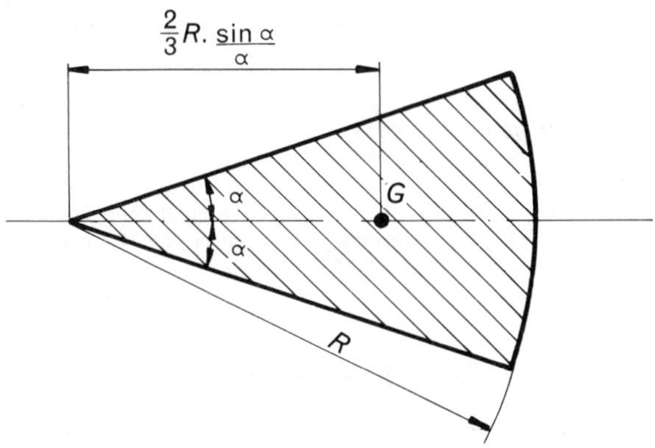

Sector of a circle

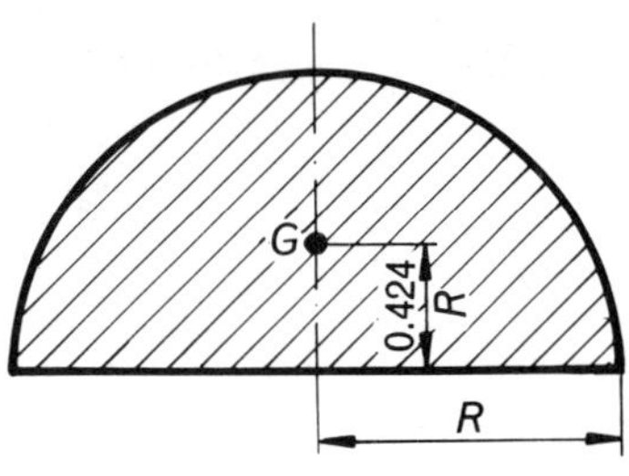

Semi-circle

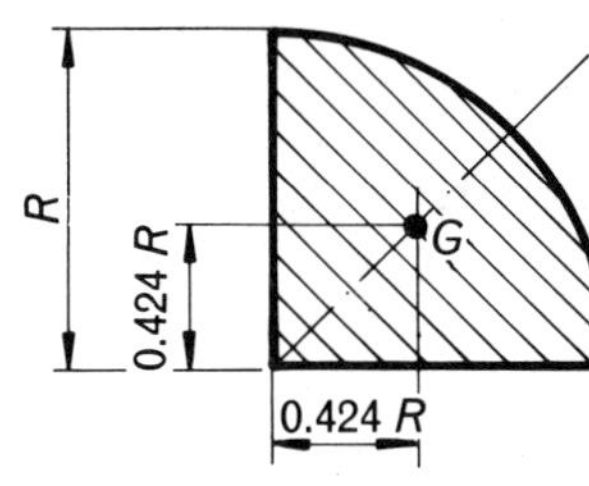

Quadrant of a circle

11. SECOND MOMENTS OF AREA

On reaching the end of this chapter you should be able to:-

1. *Sketch the given area including a typical incremental area parallel to a specified axis in the plane of the area.*
2. *Define the second moment of area.*
3. *Determine the second moment of the increment in 1 about the specified axis.*
4. *Determine the second moment of area by summing the second moment of the incremental area between given limits by definite intergation.*
5. *Determine the second moment of a rectangle about one edge, a triangle about one edge, a circle and semi-circle about a diameter, and a circle about its polar axis.*
6. *State the parallel axis theorem.*
7. *Determine the second moment of area about an axis through the centroid of the area parallel to the axis about which the second moment known using 6.*
8. *Determine the second moment of area about axis parallel to the axis about which the seco moment is known using 6 and 7.*
9. *Calculate the second moments of the co ponent areas of a composite area about common axis using 8.*
10. *Calculate the second moment of a compos area by summing the second moments in 9.*
11. *State the perpendicular axis theorem.*
12. *Calculate the second moments of area of a giv area about two axes at right angles in the pla of the given area.*
13. *Calculate the second moment of area abou perpendicular axis through the intersection axes in 12 using 11.*

SECOND MOMENT OF AREA

The second moment of area is a property of an area used in many engineering calculations. One example is in finding stresses, due to bending, which involves the use of the second moment of the cross-sectional area of the beam.

Fig. 11.1 shows a thin strip, of area A and of very small width, which is distant x from the reference line CD.

Now the first moment of the area A about the reference line CD is given by $A.x$.

Similarly the second moment of area A about the reference line CD is given by $A.x^2$.

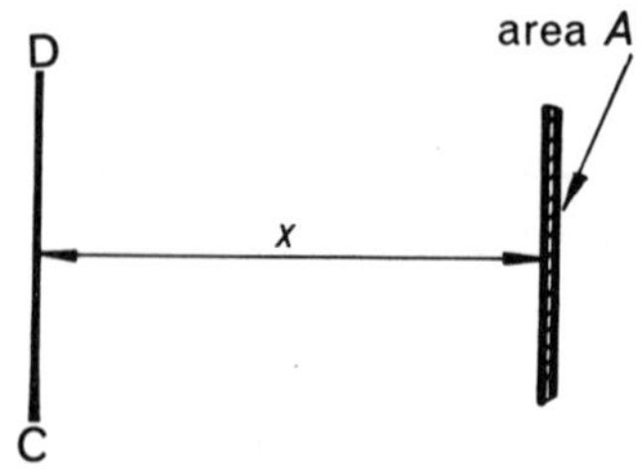

Fig. 11.1

The symbol for the second moment of area is I and is always stated with reference to an axis or datum line. Thus for the strip area shown

$$I_{CD} = A.x^2$$

The expression $A.x^2$ is only true if the area is a very thin strip parallel to the reference line. It follows that if we require the second moment of any other shaped area it is necessary to divide the area into a number of elementary strips all parallel to the reference line.

Thus in Fig. 11.2 for the irregular area $I_{CD} = \Sigma A.x^2$.

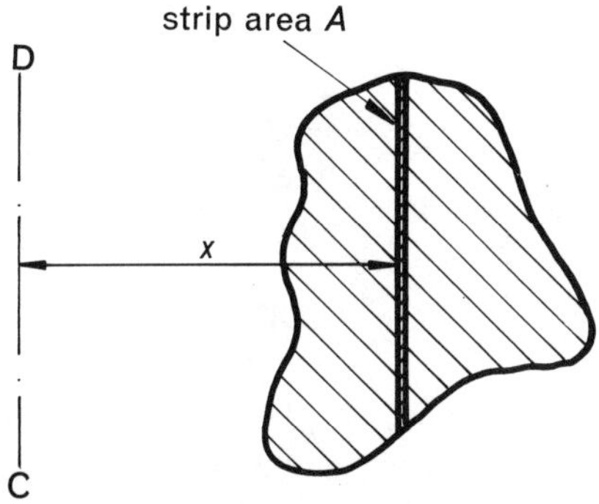

Fig. 11.2

The summation of the $A.x^2$ for each strip may be carried out by graphical and numerical methods but, as in the case of finding first moments of area, this summation may often be achieved by integration as the following example will illustrate:

EXAMPLE 1

Find the second moment of area of the rectangle shown in Fig. 11.3 about its base edge.

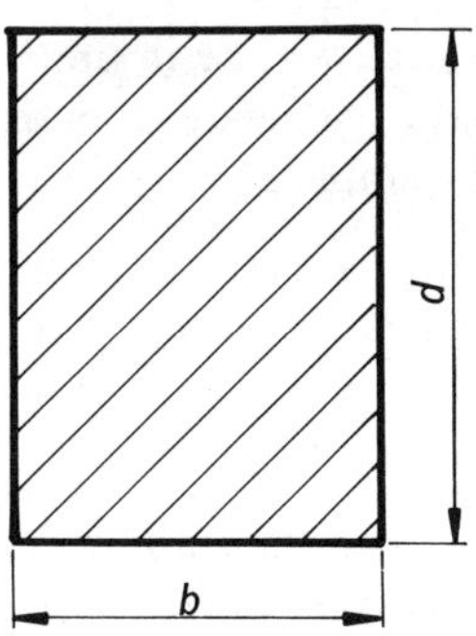

Fig. 11.3

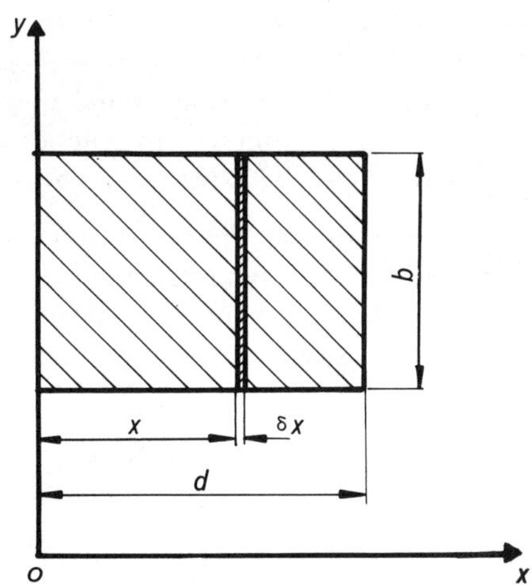

Fig. 11.4

The rectangular area must be set up on suitable axes. In this case it is convenient to turn the rectangle through 90° and arrange the base edge to lie on the y-axis as shown in Fig. 11.4. In this instance the position of the x-axis is unimportant.

The diagram shows a typical elementary strip area parallel to the reference axis and whose area is $b.\delta x$.

Now the second moment of the rectangular area:

$$\text{about the } y\text{-axis} = \Sigma\, A.x^2$$

$$= \sum_{x=0}^{x=d} b.\delta x.x^2$$

$$= \int_0^d b.\mathrm{d}x.x^2$$

$$= b\int_0^d x^2\,\mathrm{d}x$$

$$= b\left[\frac{x^3}{3}\right]_0^d = b\left\{\frac{\mathrm{d}^3}{3}-\frac{0^3}{3}\right\} = \frac{bd^3}{3}$$

UNITS OF SECOND MOMENTS OF AREA

The second moment of area is $\Sigma\, A.x^2$. If all lengths are in metres, then the result will be $\mathrm{m}^2\times\mathrm{m}^2 = \mathrm{m}^4$.

For example in the formula $\dfrac{bd^3}{3}$ the units are $\mathrm{m}\times\mathrm{m}^3 = \mathrm{m}^4$.

In typical engineering problems it is unlikely that dimensions of a cross-sectional area will be in metres. It is more probable that the units will be millimetres, and if these are used the second moment of area will be in mm^4 which will result in large numbers. In practice it is usual to calculate second moments of area in cm^4. This makes the arithmetic as simple as possible.

This is one of the few times when units not recommended by the Système International (SI) are used in engineering calculations. In British industry producers of rolled steel sections used for beams and columns, etc. agreed with continental manufacturers to use cm^4 as units of second moments of area.

The second moment of the rectangular area shown in Fig. 11.5 about its base, or more correctly about an axis BB on which the base lies, is,

$$I_{BB} = \frac{bd^3}{3} = \frac{3\times 2^3}{3} = 8\ \mathrm{cm}^4$$

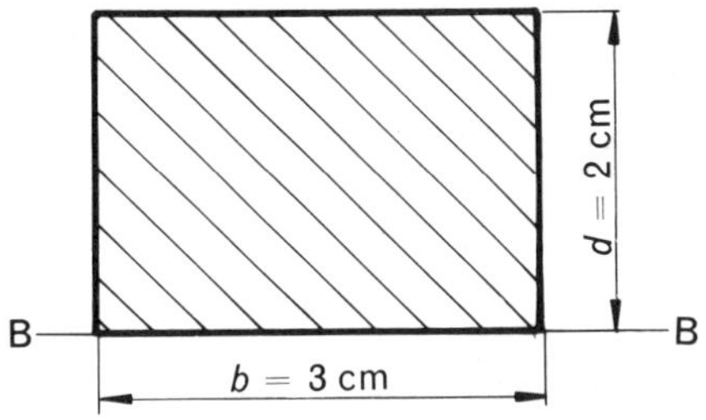

Fig. 11.5

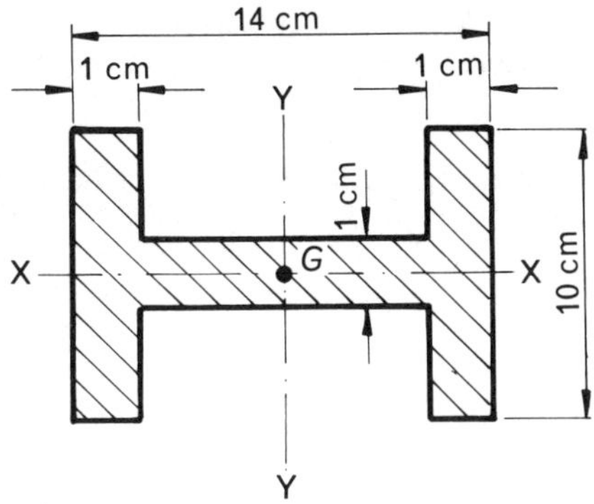

Fig. 11.6

CONVENTIONAL NOTATION I_{XX} AND I_{YY}

XX and YY are known conventionally as the horizontal and vertical axes which pass through the centroid of a cross-sectional area.

Hence: I_{XX} is the second moment of a sectional area about the horizontal axis which passes through the centroid,

and: I_{YY} is the second moment of a sectional area about the vertical axis which passes through the centroid.

EXAMPLE 2

Find I_{XX} and I_{YY} of the cross-sectional area shown in Fig. 11.6.

In this case by symmetry the centroid G will be at the centre of the section.

To find I_{XX}

We shall consider the area to be made up by six rectangles, as shown in Fig. 11.7, each of which has its 'base' on XX.

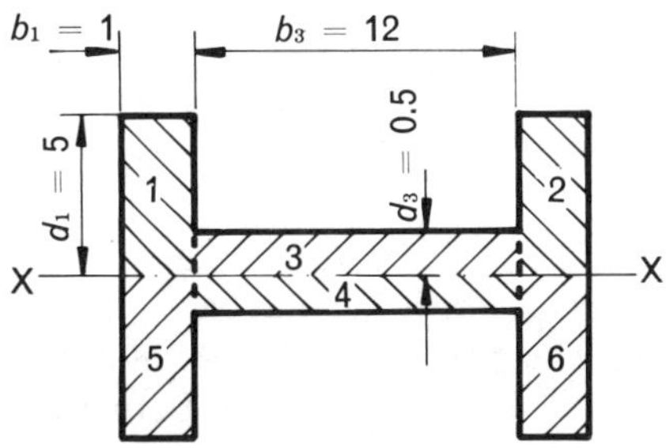

Fig. 11.7

Now I_{XX} for the cross-sectional area will be the sum of the second moments of area of each rectangle about XX.

Rectangles 1, 2, 5 and 6 have the same dimensions and hence the same I_{XX}. Similarly rectangles 3 and 4 have the same I_{XX}.

Hence I_{XX} for whole area $= 4(I_{XX}$ for area 1$)+2(I_{XX}$ for area 3$)$

$$= 4\left(\frac{b_1d_1^3}{3}\right)+2\left(\frac{b_3d_3^3}{3}\right)$$

$$= 4\left(\frac{1\times 5^3}{3}\right)+2\left(\frac{12\times 0.5^3}{3}\right)$$

$$= 166.7+1.0$$

$$= 167.7 \text{ cm}^4$$

To find I_{YY}

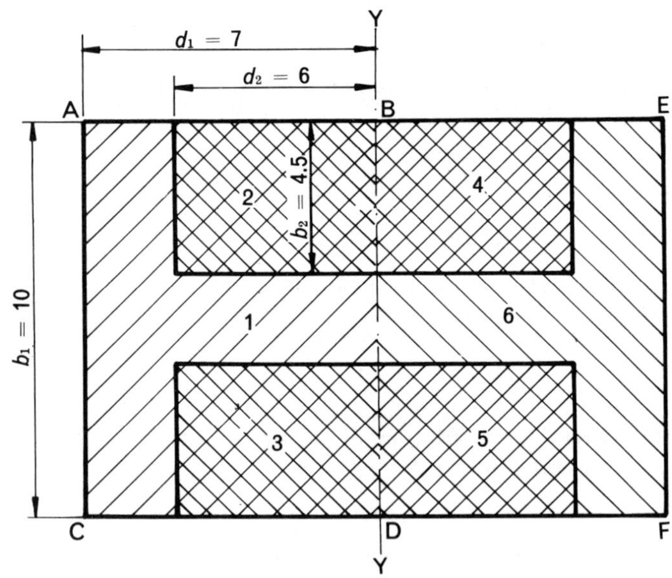

Fig. 11.8

As before we shall consider the area to be made up by six rectangles, as shown in Fig. 11.8, each of which has its 'base' on YY. In this case the 'bases' will be vertical. The portion to the left of YY is area 1 (ABDC) less areas 2 and 3. This means that the I_{YY} of both areas 2 and 3 must be taken as negative when finding I_{YY} for the whole section. We treat the portion of the cross-section to the right of YY in a similar manner.

Rectangles 1 and 6 have the same dimensions and hence the same I_{YY}. Similary rectangles 2, 3, 4 and 5 have the same I_{YY}.

Hence I_{YY} for whole area $= 2(I_{YY}$ for area 1$)-4(I_{YY}$ for area 2$)$

$$= 2\left(\frac{b_1d_1^3}{3}\right)-4\left(\frac{b_2d_2^3}{3}\right)$$

$$= 2\left(\frac{10\times 7^3}{3}\right)-4\left(\frac{4.5\times 6^3}{3}\right)$$

$$= 2287-1296$$

$$= 991 \text{ cm}^4$$

EXAMPLE 3

Find I_{XX} and I_{YY} for the cross-sectional area shown in Fig. 11.9.

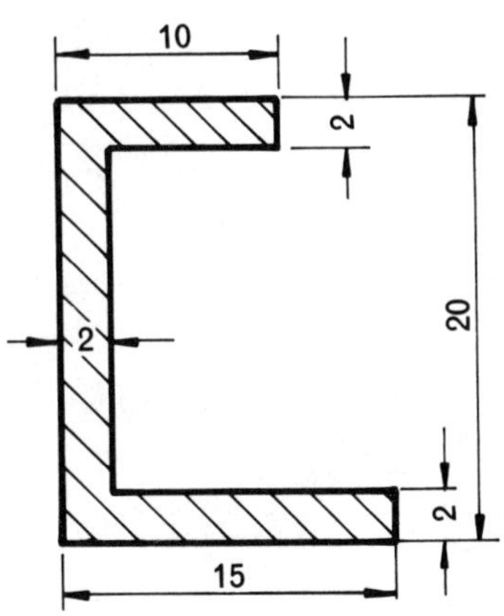

All dimensions are cm

Fig. 11.9

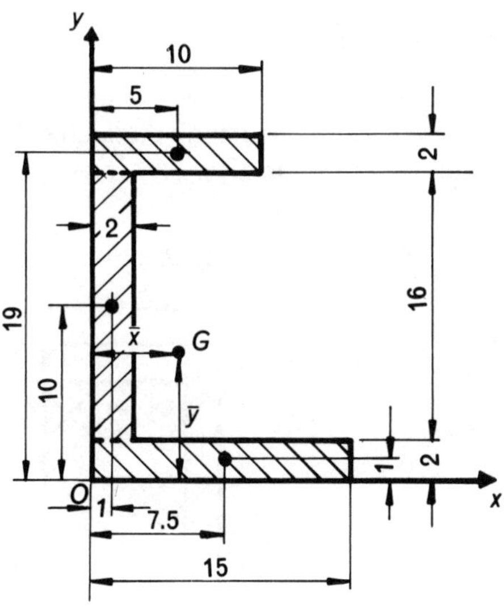

Fig. 11.10

First it is necessary to find the position of the centroid G. We will divide the cross-section into three rectangular areas as shown in Fig. 11.10 and then find the position of G relative to the chosen reference axes Ox and Oy.

Area	Distance to centroid		First moment of area	
A	from Oy x	from Ox y	about Oy Ax	about Ox Ay
$10 \times 2 = 20$	5	19	$20 \times 5 = 100$	$20 \times 19 = 380$
$2 \times 16 = 32$	1	10	$32 \times 1 = 32$	$32 \times 10 = 320$
$15 \times 2 = 30$	7.5	1	$30 \times 7.5 = 225$	$30 \times 1 = 30$
$\Sigma A = 82$			$\Sigma Ax = 357$	$\Sigma Ay = 730$

Hence $$\bar{x} = \frac{\Sigma Ax}{\Sigma A} = \frac{357}{82} = 4.35 \text{ cm}$$

and $$\bar{y} = \frac{\Sigma Ay}{\Sigma A} = \frac{730}{82} = 8.90 \text{ cm}$$

To find I_{XX}

We must now sketch a new diagram, as shown in Fig. 11.11, showing the cross-section divided into rectangles having one edge on XX (the horizontal axis through G).

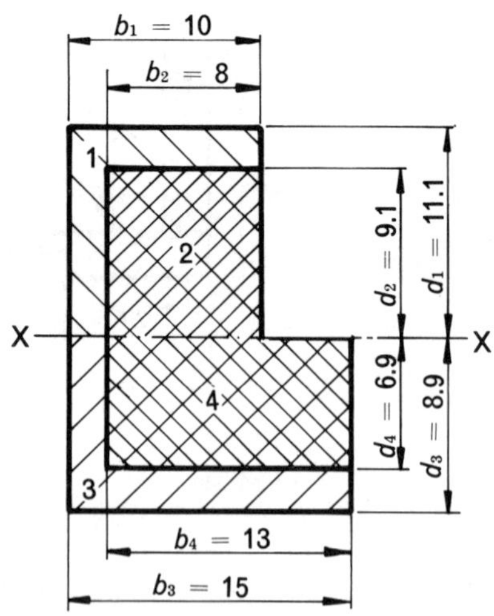

Fig. 11.11

Hence,

I_{XX} for whole area $= I_{XX}$ for area 1 $- I_{XX}$ for area 2 $+ I_{XX}$ for area 3 $- I_{XX}$ for area 4

$$= \frac{b_1 d_1^3}{3} - \frac{b_2 d_2^3}{3} + \frac{b_3 d_3^3}{3} - \frac{b_4 d_4^3}{3}$$

$$= \frac{10(11.1)^3}{3} - \frac{8(9.1)^3}{3} + \frac{15(8.9)^3}{3} - \frac{13(6.9)^3}{3}$$

$$= 4560 - 2010 + 3520 - 1420$$

$$= 4650 \text{ cm}^4$$

To find I_{YY}

Again a new sketch is required, as shown in Fig. 11.12, showing the cross-section divided into rectangles having one edge on YY (the vertical axis through G).

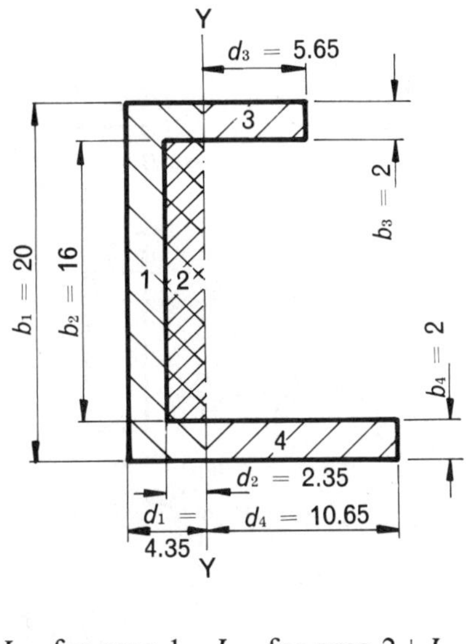

Fig. 11.12

Hence,

I_{YY} for whole area $= I_{YY}$ for area 1 $- I_{YY}$ for area 2 $+ I_{XX}$ for area 3 $+ I_{YY}$ for area 4

$$= \frac{b_1 d_1^3}{3} - \frac{b_2 d_2^3}{3} + \frac{b_3 d_3^3}{3} + \frac{b_4 d_4^3}{3}$$

$$= \frac{20(4.35)^3}{3} - \frac{16(2.35)^3}{3} + \frac{2(5.65)^3}{3} + \frac{2(10.65)^3}{3}$$

$$= 549 - 69 + 120 + 805$$

$$= 1405 \text{ cm}^4$$

Exercise 34

Find I_{XX} and I_{YY} for the cross-sectional areas given in the following examples. All dimensions are cm.

1)

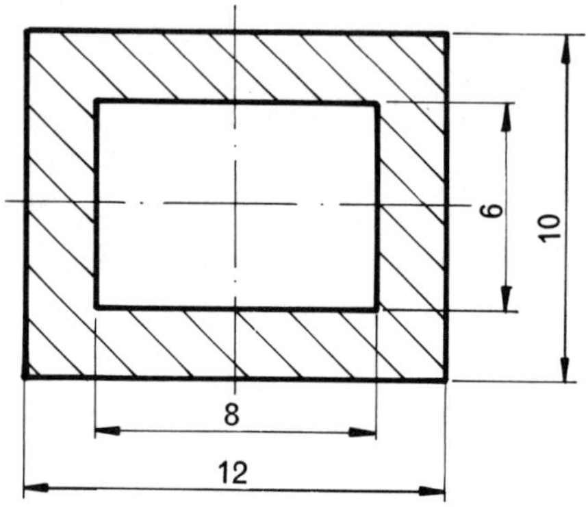

2)

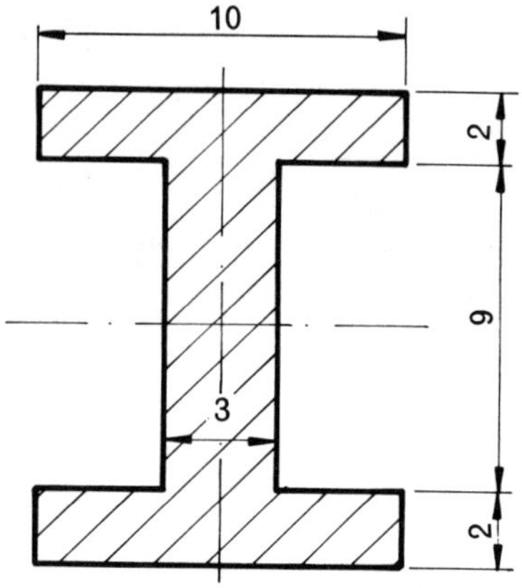

3)

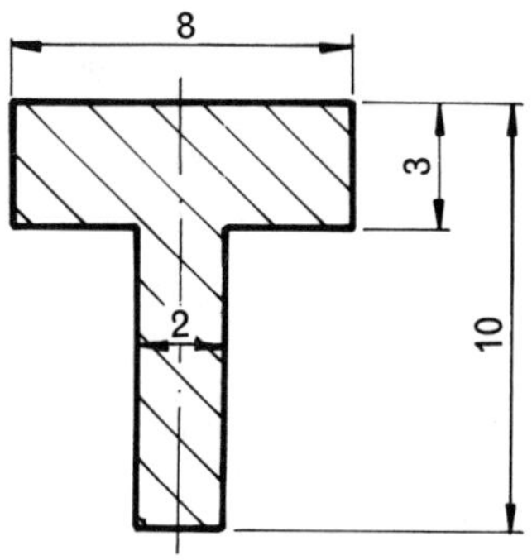

4)

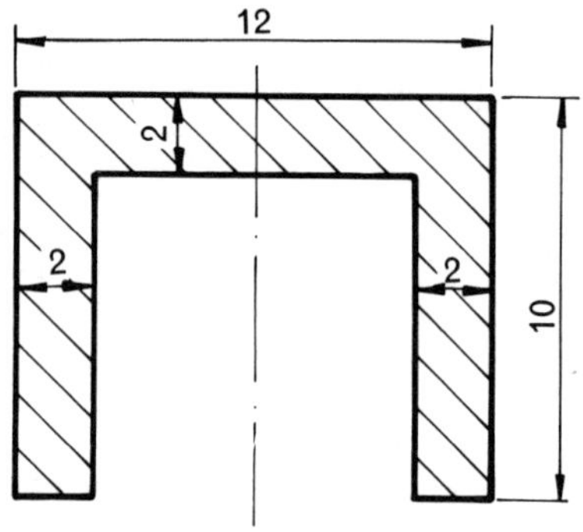

5)

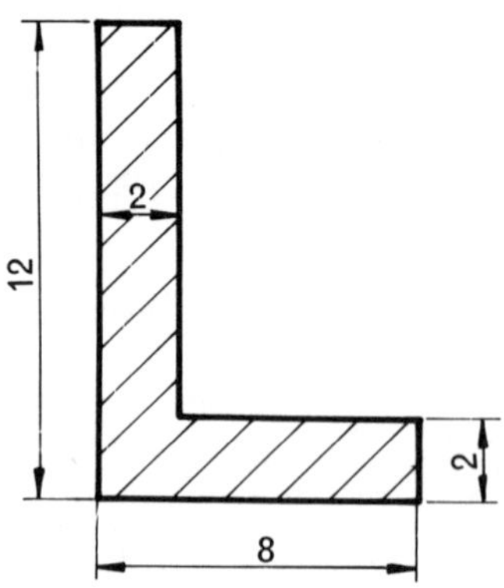

6)

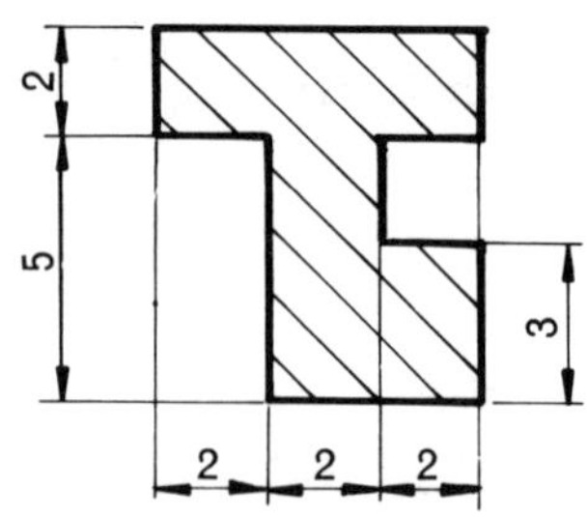

EXAMPLE 4

Find the second moment of a triangular area, of height H and base B, about a line through its apex and parallel to the base.

The arrangement of the axes etc. is identical to that used in Example 3, p. 159, and is as shown in Fig. 11.13.

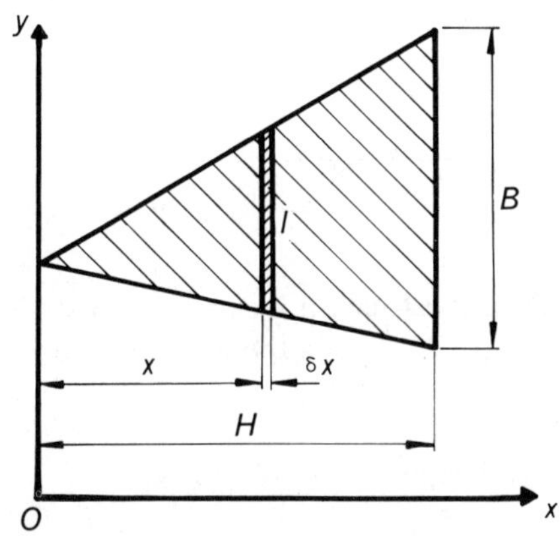

Fig. 11.13

Now the second moment of the triangular area about the y-axis:

$$= \Sigma A.x^2$$

$$= \sum_{x=0}^{x=H} l.\delta x.x^2$$

$$= \int_0^H l.x^2 \, dx$$

and using similar triangles as before we have $l = \frac{B}{H}.x$

$$\therefore \qquad I_{Oy} = \int_0^H \frac{B}{H}.x.x^2 \, dx$$

$$= \frac{B}{H}\int_0^H x^3 \, dx$$

$$= \frac{B}{H}\left[\frac{x^4}{4}\right]_0^H = \frac{B}{H}\left\{\frac{H^4}{4} - \frac{0^4}{4}\right\} = \frac{B}{H}.\frac{H^4}{4} = \frac{BH^3}{4}$$

THE PARALLEL AXIS THEOREM

A second moment of area must always be stated together with the axis about which it has been calculated. We state by saying, for example, that I_{AB} is the second moment of area about the axis AB.

Fig. 11.14 shows a plane area A whose centroid is G. Also shown is an axis which passes through G and a parallel axis, the distance between the axes being h.

The parallel axis theorem states:

$$I_{\text{parallel axis}} = I_{\text{axis through }G} + A.h^2$$

This is extremely useful as the following examples will show. It is worth while remembering that $I_{\text{axis through }G}$ is the least second moment of area.

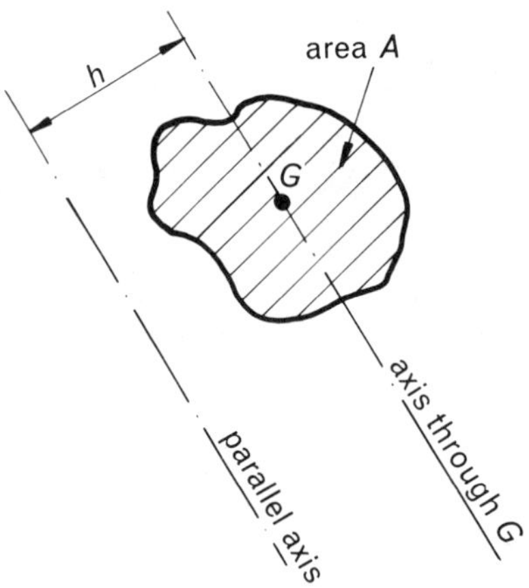

Fig. 11.14

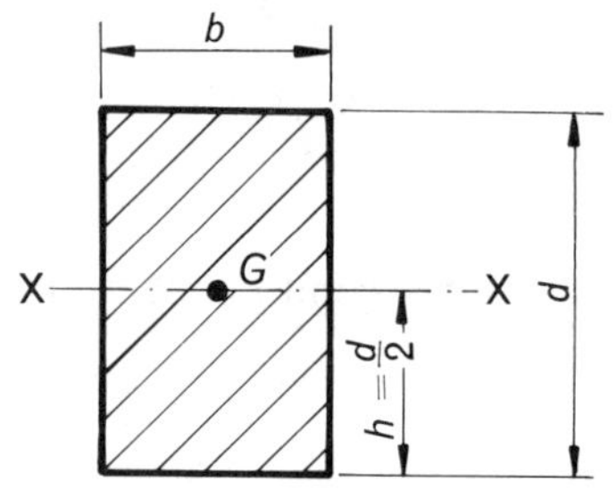

Fig. 11.15

EXAMPLE 5

Find the second moment of area of the rectangle shown in Fig. 11.15 about the axis XX which passes through the centroid.

We know from Example 1 that the second moment of area of the rectangle about its base is $\dfrac{b.d^3}{3}$.

Now using the parallel axis theorem we have

$$I_{\text{base}} = I_{XX} + A.h^2$$

$\therefore$ rearranging
$$I_{XX} = I_{\text{base}} - A.h^2$$
$$= \frac{bd^3}{3} - bd\left(\frac{d}{2}\right)^2$$
$$= bd^3\left(\frac{1}{3} - \frac{1}{4}\right)$$
$$= \frac{bd^3}{12}$$

For many of the shapes used in engineering a tabular method using the theorem of parallel axes is often used as an alternative to the method used in Examples 2 and 3.

EXAMPLE 6

Find I_{XX} for the cross-section shown in Fig. 11.16.

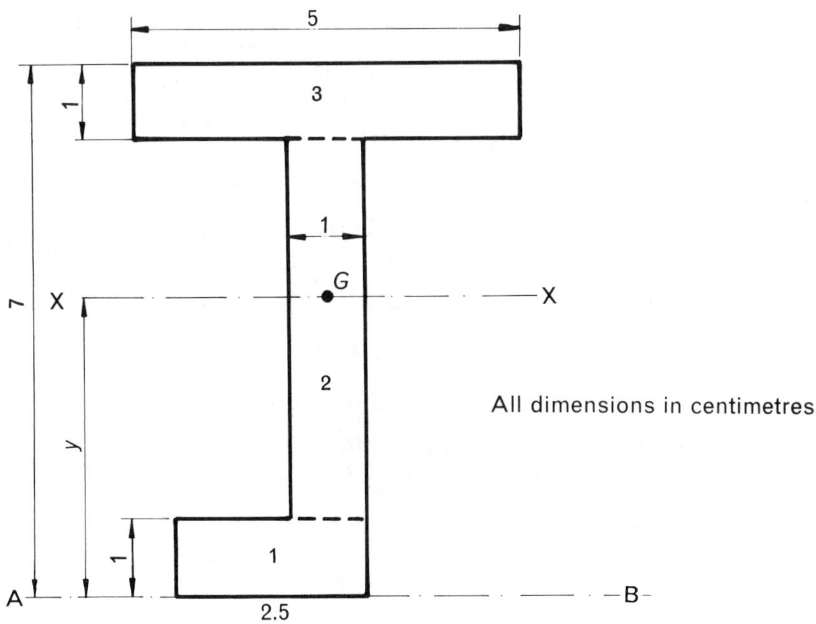

Fig. 11.16

The first step is to choose a suitable axis, such as AB in Fig. 11.16, and calculate I_{AB}. To do this the cross-section is divided into the three rectangles 1, 2 and 3. Note that for each rectangle $I_{AB} = Ay^2 + I_G$

where A = area of the rectangle

y = distance of centroid of the rectangle from AB

I_G = second moment of area of the rectangle about the centroid of the rectangle $= \dfrac{bd^3}{12}$

Item	A	y	Ay	Ay^2	$I_G = \dfrac{bd^3}{12}$
1	2.5	0.5	1.25	0.6	$\dfrac{2.5\times 1^3}{12} = 0.2$
2	5.0	3.5	17.50	61.3	$\dfrac{1\times 5^3}{12} = 10.4$
3	5.0	6.5	32.50	211.3	$\dfrac{5\times 1^3}{12} = 0.4$
	12.5		51.25	273.2	11.0

$$\bar{y} = \frac{\Sigma\, Ay}{\Sigma\, A} = \frac{51.25}{12.5} = 4.10 \text{ cm}$$

∴ total

$$I_{AB} = \Sigma\, Ay^2 + \Sigma\, I_G$$
$$= 273.2 + 11.0 = 284.2 \text{ cm}^4$$

and again using the parallel axis we have:

$$I_{XX} = I_{AB} - (\Sigma\, A)x\bar{y}^2$$
$$= 284.2 - 12.5\times(4.10)^2$$
$$= 284.2 - 210.1$$
$$= 74.1 \text{ cm}^4$$

EXAMPLE 7

Given that the second moment of a triangular area, of height H and base B, about a line through its apex and parallel to its base is $\dfrac{B.H^3}{4}$ find the second moment of the triangular area about its base.

Fig. 11.17 shows a sketch of the triangular area with axes suitably labelled. We are given I_{MM} and we require I_{NN}.

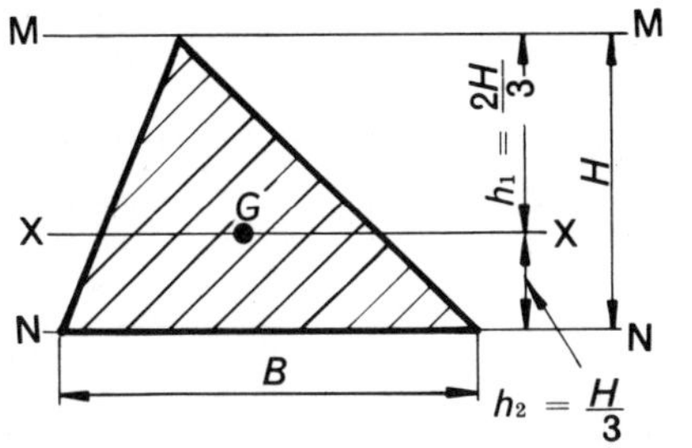

Fig. 11.17

The parallel axis theorem requires that one of the axes used must pass through the centroid. It is necessary, therefore, to use the parallel axis theorem twice — once to find I_{XX} knowing I_{MM} and again to find I_{NN} using I_{XX}.

Now the parallel axis theorem gives $\quad I_{MM} = I_{XX} + A.h^2$

and rearranging $\quad I_{XX} = I_{MM} - A.h^2{}_1$

$$\therefore \quad I_{XX} = \frac{B.H^3}{4} - \frac{B.H}{2}\left(\frac{2H}{3}\right)^2$$

$$= \frac{B.H^3}{36}$$

Also the parallel axis theorem gives $\quad I_{NN} = I_{XX} + A.h_2{}^2$

$$= \frac{B.H^3}{36} + \frac{B.H}{2}\left(\frac{H}{3}\right)^2$$

$$= \frac{B.H^3}{12}$$

Hence the second moment of a triangular area about its base is $\dfrac{B.H^3}{12}$.

EXAMPLE 8

The second moment of a rectangular area, of base length b and depth d, about an axis through its centroid and parallel to the base is $\dfrac{b.d^3}{12}$. The second moment of a triangular area, of height H and base length B, about an axis along its base is $\dfrac{B.H^3}{12}$.

Using the above data and also the parallel axis theorem for the area shown in Fig. 11.18, find:

(a) I_{XX} (b) I_{YY} (c) I_{CD}

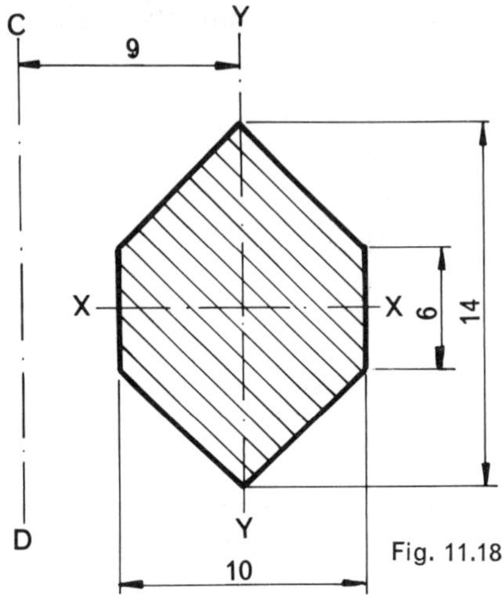

Fig. 11.18

All dimensions are cm

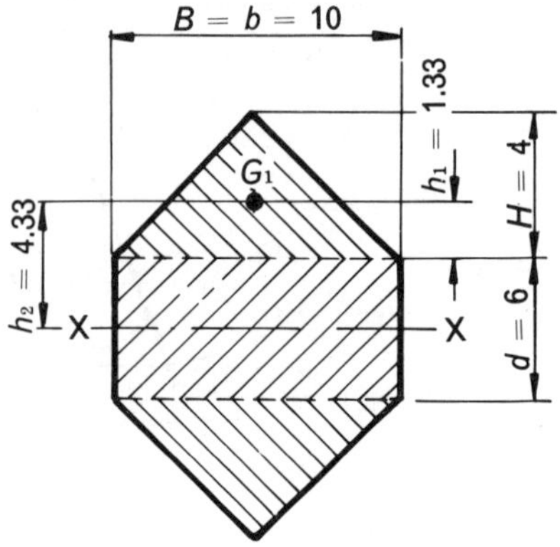

Fig. 11.19

(a) To find I_{XX}

Fig. 11.19 shows the given area divided into a rectangle and two triangles. Since the triangles are of identical dimensions and the same disposition relative to the axis XX then their second moments of area about XX will be the same. The parallel axis theorem will be used to find I_{XX} of the triangular areas about XX knowing their I_{base}. This will have to be done in two stages:

(i) To find I of the triangle about a line parallel to the base of the triangle passing through G_T (G_T being the centroid of the triangle).

(ii) To find I_{XX} for the triangle.

For the triangular area the parallel axis theorem gives

$$I_{base} = I_{G_T} + (\text{area of triangle}) \times h_1^2$$

$\therefore$ rearranging

$$I_{G_T} = I_{base} - (\text{area of triangle}) \times h_1^2$$

$$= \frac{B.H^3}{12} - \frac{1}{2}.B.H \times h_1^2$$

$$= \frac{10 \times 4^3}{12} - \frac{1}{2} \times 10 \times 4 \times 1.33^2$$

$$= 53.3 - 35.4$$

$$= 17.9 \text{ cm}^4$$

For the same triangular area the parallel axis theorem gives

$$I_{XX} = I_{G_T} + (\text{area of triangle}) \times h_2^2$$

$$= 17.9 + \left(\frac{1}{2} \times 10 \times 4\right) \times 4.33^2$$

$$= 17.9 + 375$$

$$= 393 \text{ cm}^4$$

Now for the given area

$$I_{XX} \text{ of given area} = I_{XX} \text{ of rectangle} + 2(I_{XX} \text{ of triangle})$$

$$= \frac{b.d^3}{12} + 2(393)$$

$$= \frac{10 \times 6^3}{12} + 786$$

$$= 180 + 786$$

$$= 966 \text{ cm}^4$$

(b) To find I_{YY}

Fig. 11.20 shows the given area divided into a rectangle and four triangles of equal dimensions.

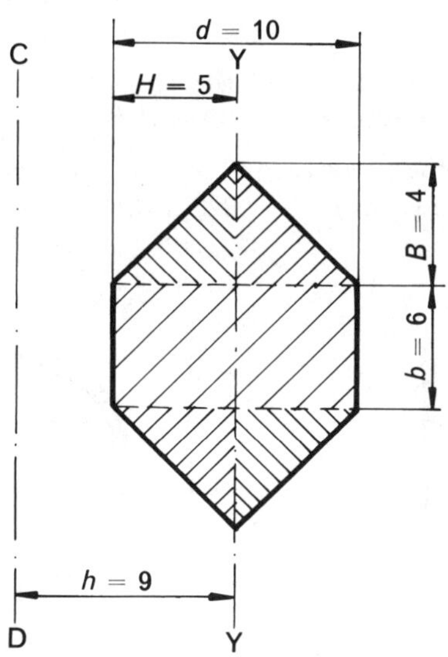

Fig. 11.20

Hence I_{YY} of given area $= I_{YY}$ of rectangle$+4(I_{YY}$ of triangle)

$$= \frac{b.d^3}{12}+4\left(\frac{B.H^3}{12}\right)$$

$$= \frac{6\times10^3}{12}+4\left(\frac{4\times5^3}{12}\right)$$

$$= 500+167$$

$$= 667 \text{ cm}^4$$

(c) I_{CD} may be found using the parallel axis theorem for the given area.

Hence I_{CD} of given area $= I_{YY}$ of given area$+$(given area)$\times h^2$

$$= 667+\left(6\times10+2\times\frac{1}{2}\times10\times4\right)\times9^2$$

$$= 667+8100$$

$$= 8767 \text{ cm}^4$$

POLAR SECOND MOMENT OF AREA

The polar second moment of area, denoted by the symbol J, is the second moment of a circular area about the polar axis (that is the axis passing through the centre of the area and perpendicular to the plane of the area).

One example of its use is in finding stresses due to torsion in a circular shaft.

Now the second moment of area of the elementary strip area, shown in Fig. 11.21, about the reference line CD is $A.x^2$.

Similarly we say that the polar second moment of area of the elementary circular strip area, shown in Fig. 11.22, is $A.r^2$.

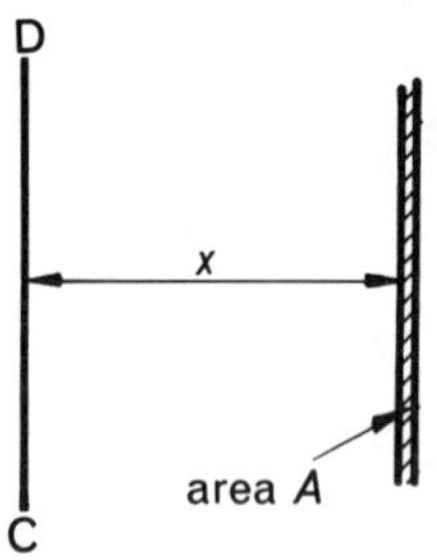

Fig. 11.21

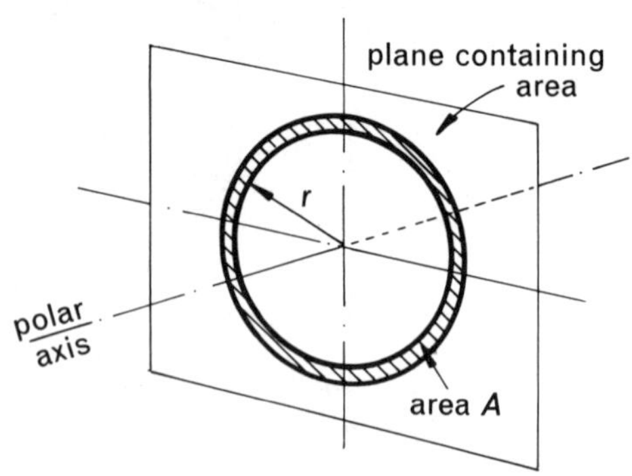

Fig. 11.22

EXAMPLE 9

Find the polar second moment of a circular area of diameter D.

Fig. 11.23 shows the given area and an elementary circular strip. We shall sum the $A.x^2$ for each strip area to find the polar second moment of area for the whole area.

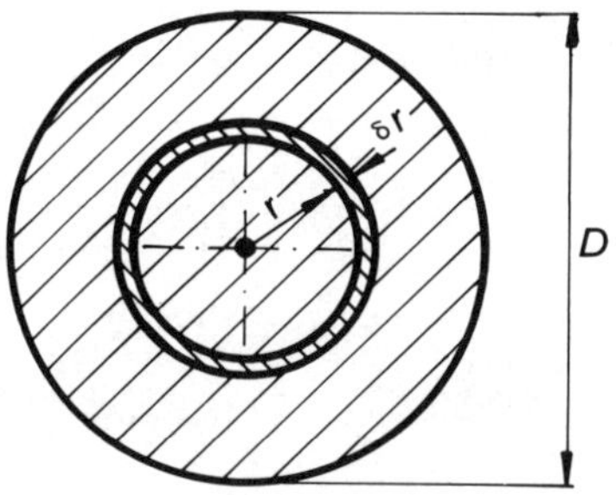

Fig. 11.23

Hence J for the circular area $= \Sigma\, A.r^2$

Now the approximate area A of the elementary circular strip

$$= \text{(circumference of strip)} \times \text{(width of strip)}$$

$$= 2\pi r \times \delta r$$

$$\therefore \quad J \text{ for the circular area} = \sum_{r=0}^{r=D/2} 2\pi r.\delta r.r^2$$

$$= \int_0^{D/2} 2\pi r.\mathrm{d}r.r^2$$

$$= 2\pi \int_0^{D/2} r^3\,\mathrm{d}r$$

$$= 2\pi \left[\frac{r^4}{4}\right]_0^{D/2}$$

$$= 2\pi \left\{\frac{(D/2)^4}{4} - \frac{0^4}{4}\right\}$$

$$= \frac{\pi D^4}{32}$$

EXAMPLE 10

Find the polar second moment of area of the cross-section of a tube 6 cm outside diameter and 4 cm inside diameter.

We shall consider J of the cross-sectional area shown in Fig. 11.24 to be J of a 6 cm diameter circular area less J of a 4 cm diameter circular area.

$$\therefore \qquad \text{required } J = \frac{\pi D^4}{32} - \frac{\pi d^4}{32}$$

$$= \frac{\pi 6^4}{32} - \frac{\pi 4^4}{32}$$

$$= 127 - 25$$

$$= 102 \text{ cm}^4$$

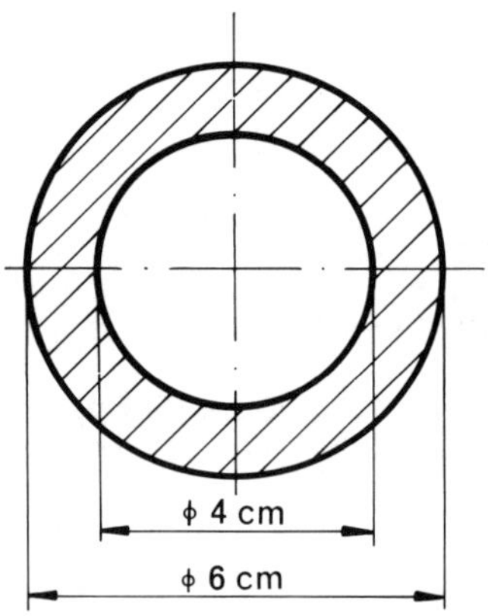

Fig. 11.24

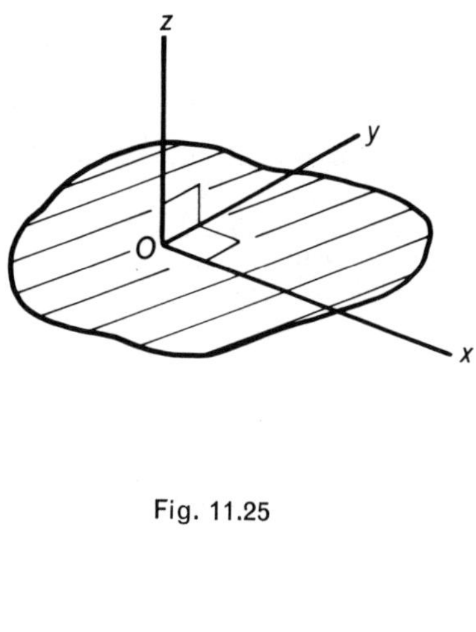

Fig. 11.25

THE PERPENDICULAR AXIS THEOREM

Fig. 11.25 shows a plane area on which the axes Ox and Oy are drawn, the axes being at right angles.

Oz is an axis perpendicular to the plane area. Ox, Oy and Oz are said to be mutually perpendicular.

The perpendicular axis theorem states:

$$I_{Oz} = I_{Ox} + I_{Oy}$$

EXAMPLE 11

Find. I_{Oz} for the rectangle shown in Fig. 11.26.

We know that for a rectangle

$$I_{\text{base}} = \frac{b.d^3}{3}$$

$$\therefore \qquad I_{\text{OA}} = \frac{12\times 8^3}{3} = 2048 \text{ cm}^4$$

$$\text{and} \qquad I_{\text{OB}} = \frac{8\times 12^3}{3} = 4608 \text{ cm}^4$$

Now the perpendicular axis theorem states:

$$\begin{aligned} I_{Oz} &= I_{\text{OA}}+I_{\text{OB}} \\ &= 2048+4608 \\ &= 6656 \text{ cm}^4 \end{aligned}$$

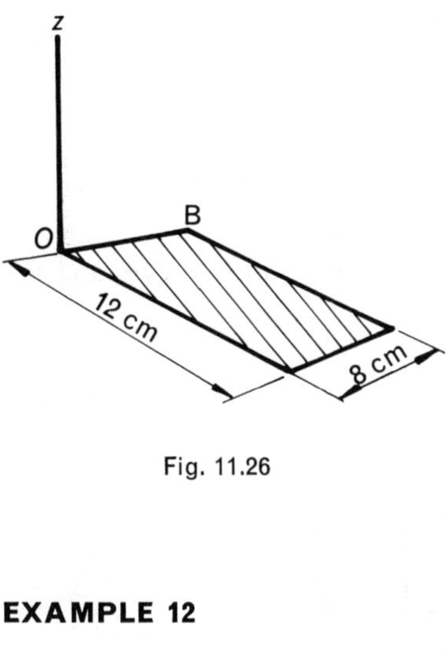

Fig. 11.26

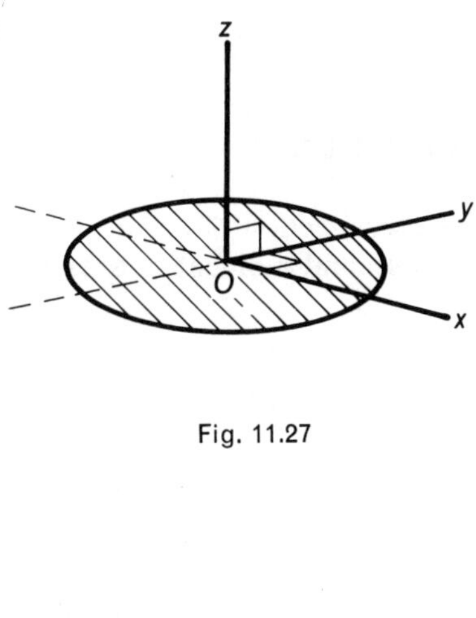

Fig. 11.27

EXAMPLE 12

Using the fact that J for a circular area is $\dfrac{\pi d^4}{32}$ and the perpendicular axis theorem find I about a diameter.

Now for a circular area I_{Oz} is called J (Fig. 11.27), and both I_{Ox} and I_{Oy} are equal to I_{diameter}.

The perpendicular axis theorem states:

$$I_{Oz} = I_{Ox}+I_{Oy}$$

$$\therefore \qquad J = I_{\text{diameter}}+I_{\text{diameter}}$$

hence
$$I_{\text{diameter}} = \frac{1}{2} . J$$
$$= \frac{1}{2} . \frac{\pi d^4}{32}$$
$$= \frac{\pi d^4}{64}$$

Hence I about a diameter of a circular area is $\frac{\pi d^4}{64}$.

A semi-circular area has I about the boundary diameter equal to one half of the result obtained for a full circular area about a diameter, hence for a semi-circular area:

$$I_{\text{diameter}} = \frac{\pi d^4}{128}$$

EXAMPLE 13

The cross-section of a shaft is shown in Fig. 11.28. Find I_{XX}, I_{YY}, and the polar second moment of area J.

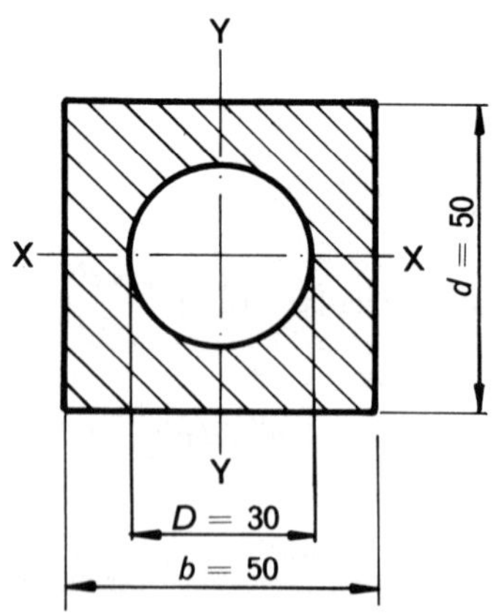

Fig. 11.28

To find I_{XX}

$$I_{XX} \text{ of cross-section} = I_{XX} \text{ of square} - I_{XX} \text{ of circle}$$

$$= \frac{b.d^3}{12} - \frac{\pi D^4}{64}$$

$$= \frac{50 \times 50^3}{12} - \frac{\pi 30^4}{64}$$

$$= 520\,800 - 39\,800$$

$$= 481\,000 \text{ mm}^4$$

$$= \frac{481\,000}{10^4} \text{ cm}^4 = 48.1 \text{ cm}^4$$

To find I_{YY}

We may see from the symmetry of the figure that $I_{YY} = I_{XX}$

Hence $I_{YY} = 48.1$ cm^4.

To find J

Using the perpendicular axis theorem we have:

$$J = I_{XX} + I_{YY}$$

$$= 48.1 + 48.1$$

$$= 96.2 \text{ cm}^4$$

RADIUS OF GYRATION

The formula for finding centroids is:

$$(\Sigma\, A).\bar{x} = \Sigma\, A.x$$

Similarly for second moments of area we have:

$$I = (\Sigma\, A).k^2 = \Sigma\, A.x^2$$

where k is called the 'radius of gyration' of the area.

In this chapter we have concentrated on using I found from $\Sigma\, A.x^2$. However we are given sometimes the area, $\Sigma\, A$, of a cross-section together with the radius of gyration, k, and we then find the second moment of area, I, by using $I = (\Sigma\, A).k^2$.

EXAMPLE 14

Find I_{XX} for a cross-sectional area given that the area of the section is 85 cm² and $k_{XX} = 3.7$ cm.

We have $$I = (\text{area of section}).k^2$$

$$\therefore \quad I_{XX} = 85(3.7)^2$$

$$= 1164 \text{ cm}^4$$

Exercise 35

Find I_{XX} and I_{YY} for the cross-sectional areas in Questions 1, 2, 3 and 4. All dimensions are cm.

1)

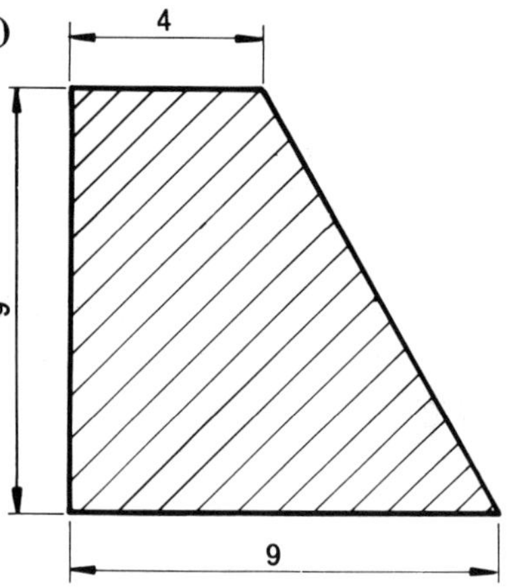

2)

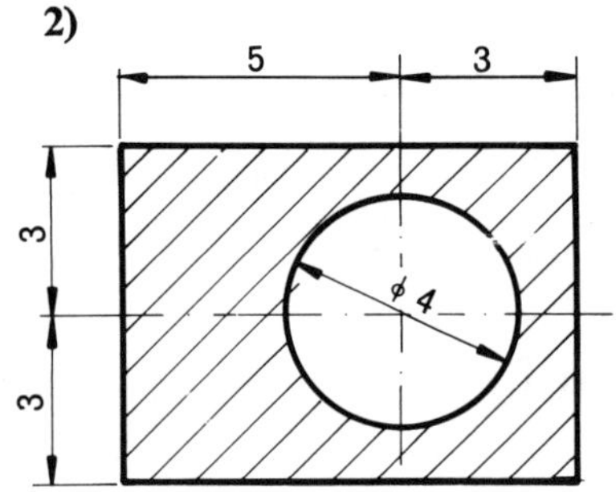

3)

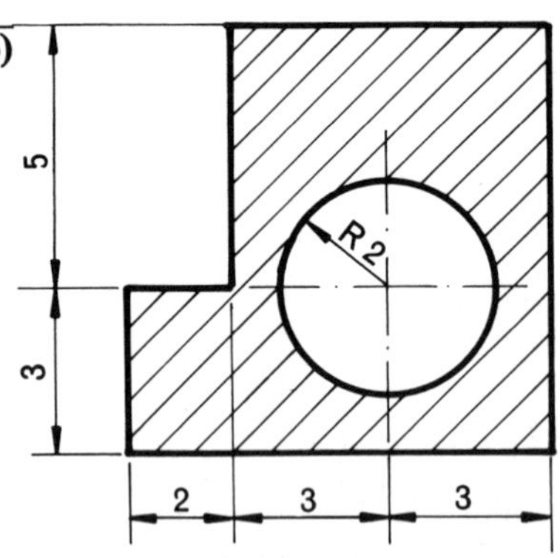

4)

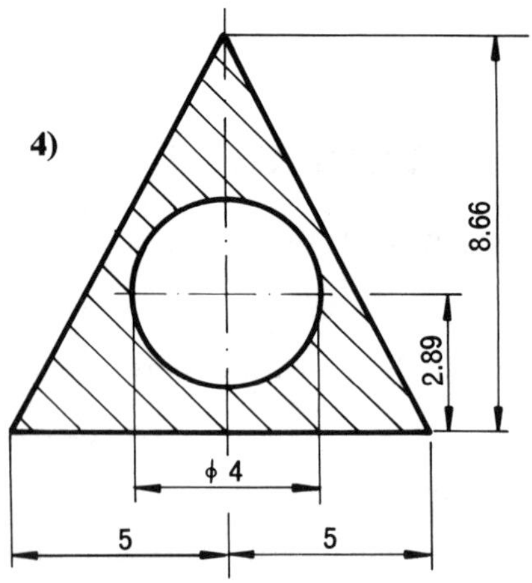

5) Using a tabular method find I_{XX} for the sections shown in Questions 2, 3, 4 and 5 of Exercise 34.

6) For the sectional area shown in Fig. 11.29 find:

(a) I about a diameter,

(b) J about the polar axis,

(c) I_{AA}.

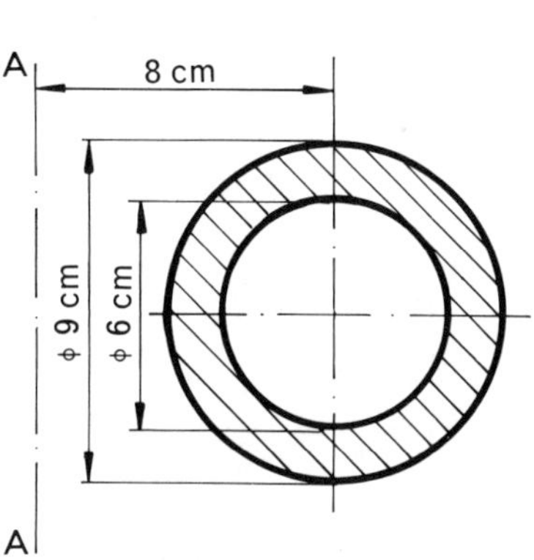

Fig. 11.29

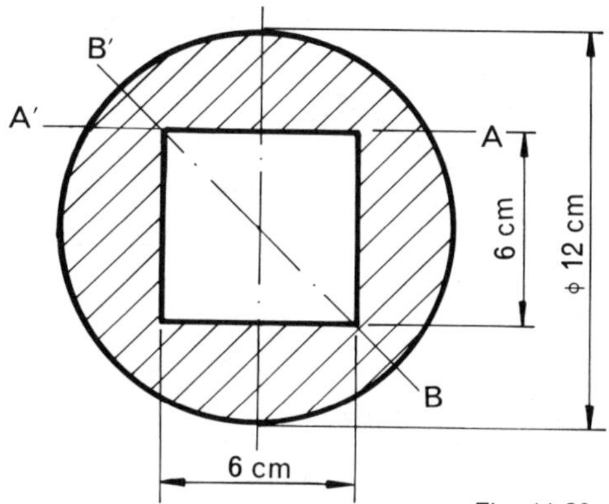

Fig. 11.30

7) Starting with the knowledge that the second moment of area of a rectangle about an edge is $b.d^3/3$ and that the second moment of area of a circle about a diameter is $\pi d^4/64$, and also using the perpendicular and parallel axis theorems, find the second moment of the area shown in Fig. 11.30 about the:

(a) axis A′A,

(b) axis through G, perpendicular to the plane of the area,

(c) axis B′B.

SUMMARY

Second moments of the more common areas:

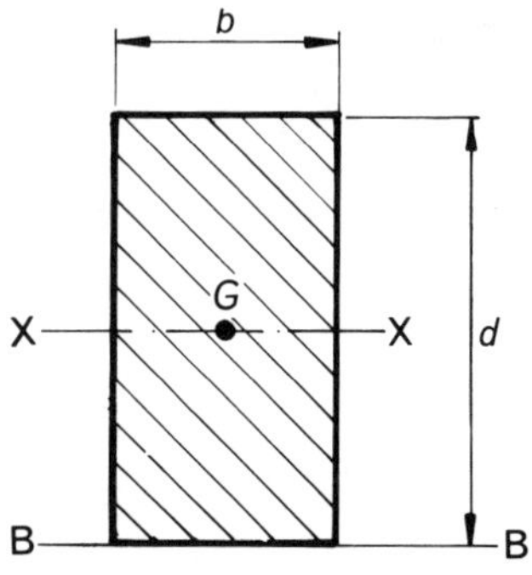

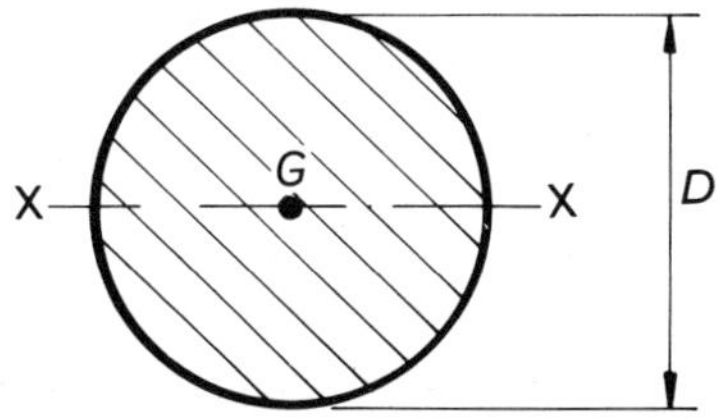

$$I_{XX} = \frac{b.d^3}{12} \quad I_{BB} = \frac{b.d^3}{3} \qquad I_{XX} = \frac{D^4}{64} \quad J = \frac{D^4}{32}$$

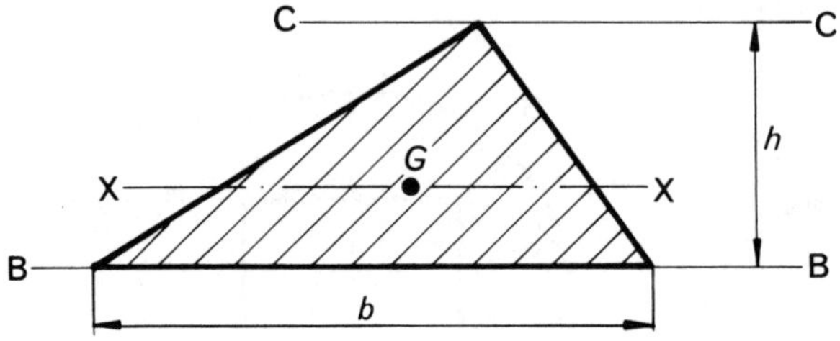

$$I_{XX} = \frac{b.h^3}{36}$$

$$I_{BB} = \frac{b.h^3}{12}$$

$$I_{CC} = \frac{b.h^3}{4}$$

Parallel Axis theorem

$$I_{BB} = I_{XX} + A.h^2$$

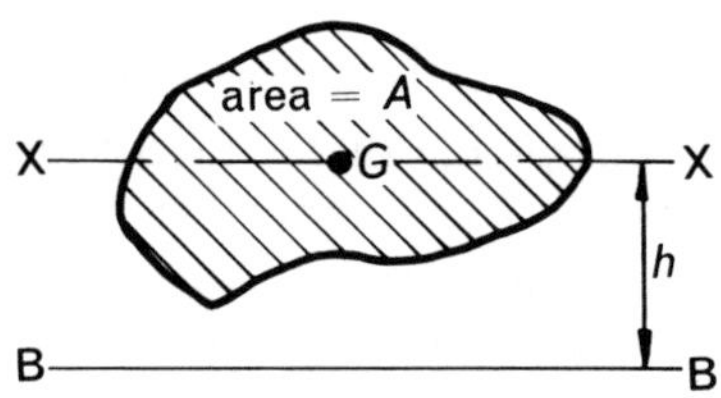

Perpendicular axis theorem

$$I_{ZZ} = I_{XX} + I_{YY}$$

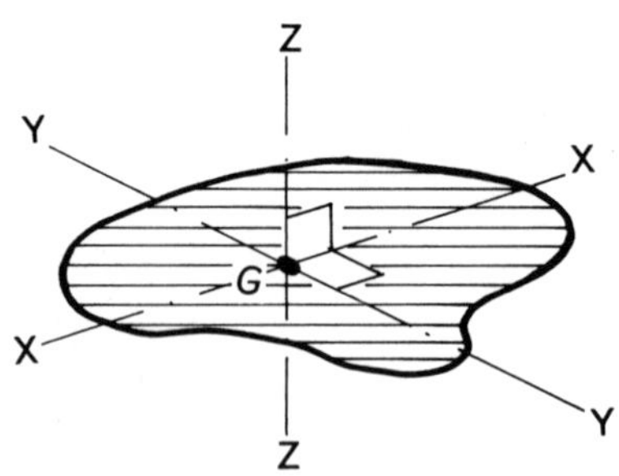

12. MOMENTS OF INERTIA

reaching the end of this chapter you should be able to:-

Determine that the moment of inertia $I = \Sigma mr^2$ *for a system of concentrated rotating masses.*
Define radius of gyration k *from the expression* $I = (\Sigma m)k^2$.
Deduce units of moment of inertia.
Understand that all the mass in an elementary ring may be considered to be concentrated at one radius — hence its polar $I = Mr^2$.

5. *Use the above information to find by integration the polar* I *of a solid cylinder.*
6. *State the parallel axis theorem.*
7. *Apply the above to typical engineering component shapes.*

MOMENT OF INERTIA

Moment of inertia is the property of a body used in rotational problems just as the mass is used in problems involving linear motion.

The symbol for moment of inertia is I and must always be stated together with the reference axis about which it has been calculated, in a similar manner as was used for second moment of area. It is unfortunate that engineers have chosen the same symbol for moment of inertia as for second moment of area but, in practice, confusion hardly ever occurs.

The linear kinetic energy of a mass m moving with a velocity v is given by $\frac{1}{2}mv^2$.

Also, the tangential velocity v at the end of a radius r rotating about an axis through O, and perpendicular to the plane of the paper, with angular velocity ω (Fig. 12.1) is given by $v = r\omega$.

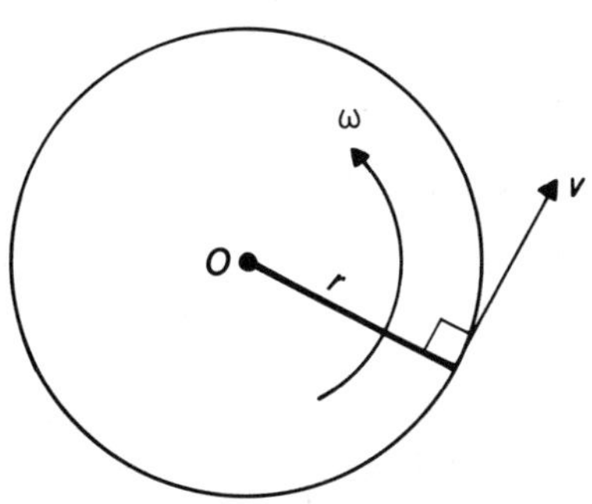

A plan view

Fig. 12.1

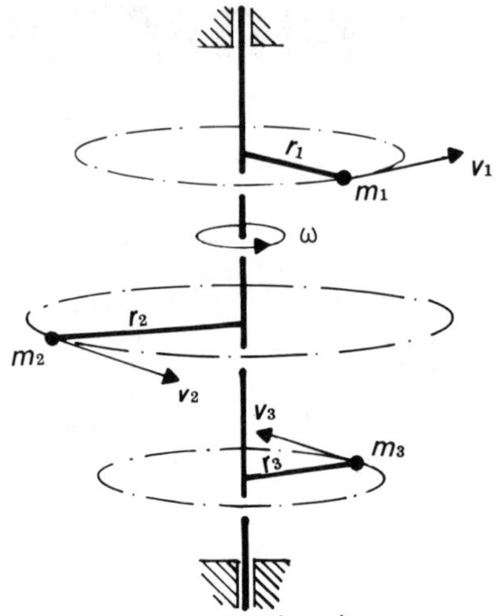

A perspective view

Fig. 12.2

Consider the system shown in Fig. 12.2 which comprises three concentrated masses m_1, m_2, and m_3 each being fixed to the end of a radius arm. The radius arms are all attached to a vertical spindle, the arms and spindle being assumed to have negligible mass. The whole system rotates with an angular velocity ω.

$$
\begin{aligned}
\text{The kinetic energy of the system} &= \frac{1}{2}m_1v_1^2+\frac{1}{2}m_2v_2^2+\frac{1}{2}m_3v_3^2\\
&= \frac{1}{2}m_1(r_1\omega)^2+\frac{1}{2}m_2(r_2\omega)^2+\frac{1}{2}m_3(r_3\omega)^2\\
&= \frac{1}{2}(m_1r_1^2+m_2r_2^2+m_3r_3^2)\omega^2\\
&= \frac{1}{2}\left(\sum mr^2\right)\omega^2\\
&= \frac{1}{2}I\omega^2
\end{aligned}
$$

where **the moment of inertia of the system is** $I = \Sigma\, mr^2$.

Although the expression $I = \Sigma\, mr^2$ has been derived for three rotating masses it is true for a system comprising any number of concentrated masses.

This expression for I may be used, together with integration, for finding the moment of inertia of bodies whose mass is *not* concentrated at any one particular radius. Example 3 shows this method.

Now for first moments of area $(\Sigma A)\bar{x} = (\Sigma Ax)$

and for second moments of area $I = (\Sigma A)k^2 = (\Sigma Ax^2)$

Similarly **for moments of inertia:** $\boxed{I = (\Sigma m)k^2 = (\Sigma mr^2)}$

where Σm is the total mass of the body

and k is the radius of gyration

UNITS OF MOMENT OF INERTIA

Since $I = (\Sigma m)k^2$ the units will be those of mass $\times$ distance². In basic SI units the mass will be kilogrammes and the distance metres, and so the units of moment of inertia will be **kg m²**.

EXAMPLE 1

Find the moment of inertia of a component about an axis, given that its mass if 5 kg and the radius of gyration about this axis is 200 mm.

Now
$$I = \text{total mass} \times k^2$$
$$= 5 \times \left(\frac{200}{1000}\right)^2$$
$$= 0.2 \text{ kg m}^2$$

EXAMPLE 2

Find the polar moment of inertia of a rim type flywheel whose dimensions are as shown in Fig. 12.3. The density of the steel from which it is made is 7800 kg/m³.

A rim type flywheel is a wheel in which all of its mass may be considered to be concentrated round the rim, the hub and spokes being neglected in calculations.

Hence all the mass may be considered to be situated at one particular radius, namely the mean radius of the rim and this may be taken as the radius of gyration.

Hence polar $k = 0.6 - 0.04 = 0.56$ m

The volume of the rim $= 2\pi \times (\text{mean radius}) \times (\text{width}) \times (\text{thickness})$
$$= 2\pi \times 0.56 \times 0.25 \times 0.08$$
$$= 0.0704 \text{ m}^3$$

and mass of the rim $=$ volume $\times$ density

$$= 0.0704 \times 7800$$

$$= 549 \text{ kg}$$

Hence polar $I =$ total mass $\times k^2$

$$= 549 \times 0.56^2$$

$$= 172 \text{ kg m}^2$$

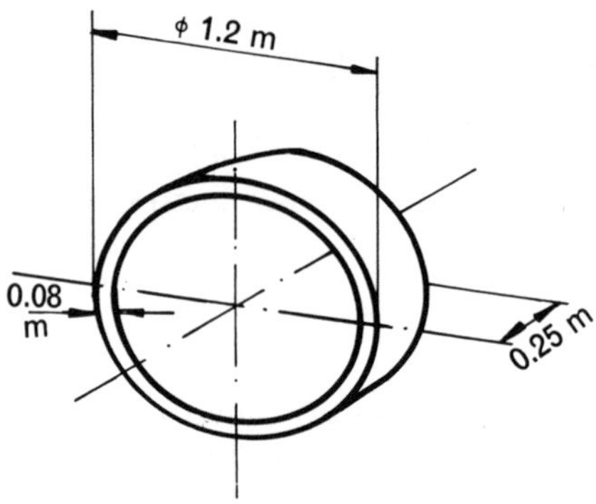

Fig. 12.3

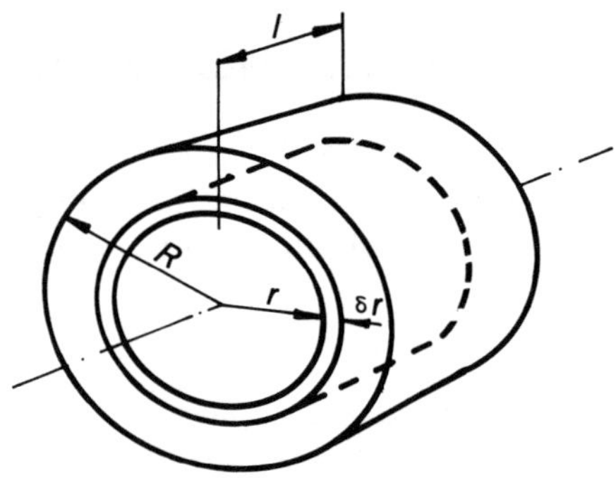

Fig. 12.4

EXAMPLE 3

Find the moment of inertia, about its polar axis, of a solid cylinder of mass M and radius R.

The mass of a solid cylinder is *not* concentrated at a particular radius and so we will consider that the cylinder is made up from a series of elementary rings (each similar to the rim of a rim type flywheel), Fig. 12.4.

Let the density of the material be ρ, and the length of the cylinder be l.

Now for the cylinder $I = \Sigma\,(mr^2)$

$$= \Sigma\,(\text{mass of elementary ring} \times r^2)$$

$$= \Sigma\,(\text{volume of elementary ring} \times \text{density} \times r^2)$$

$$= \Sigma(2\pi r\,\delta r\,l\rho r^2)$$

$$= \Sigma\,(2\pi\rho l r^3\,\delta r)$$

$$= \int_0^R 2\pi\rho l r^3\,\mathrm{d}r$$

$$= 2\pi\rho l \int_0^R r^3\,\mathrm{d}r$$

$$= 2\pi \rho l \left[\frac{r^4}{4}\right]_0^R$$

$$= \frac{2\pi \rho l R^4}{4}$$

$$= \frac{1}{2}(\pi R^2 l \rho)R^2$$

$$= \frac{1}{2}MR^2$$

since the mass of the cylinder is $M = \pi R^2 l \rho$.

If we also wished to find the radius of gyration k about the polar axis we may use: $I = Mk^2$ since $M = \Sigma m$

$\therefore$ $$\frac{1}{2}MR^2 = Mk^2$$

from which $$k^2 = \frac{R^2}{2}$$

or $$k = \frac{R}{\sqrt{2}}$$

PARALLEL AXIS THEOREM

This is similar to that for second moments of area.

In Fig. 12.5 the axis CD is parallel to the axis through centre of gravity G, h being the distance between the axes. Then:

$$I_{CD} = I_{\text{axis thro' } G} + Mh^2$$

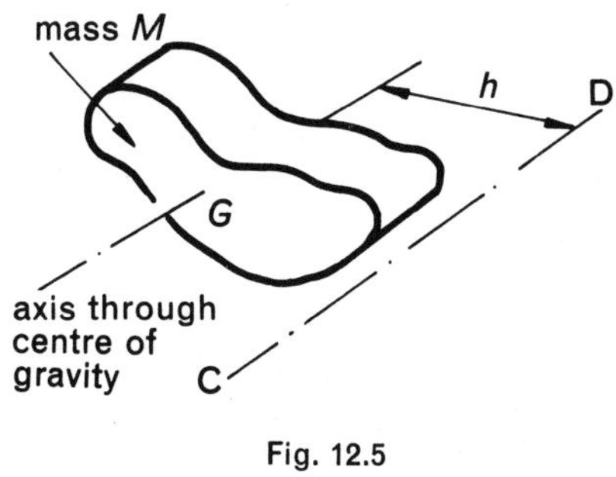

Fig. 12.5

EXAMPLE 4

Find the moment of inertia for the connecting rod shown in Fig. 12.6, about an axis through its centre of gravity G and parallel to the knife edge. Its mass is 41 kg and I about the knife edge is 15.5 kg m² found by oscillating the connecting-rod as a compound pendulum.

Using the parallel axis theorem:

$$I_{\text{knife edge}} = I_{\text{axis thro'}G} + Mh^2$$

$$\therefore \quad 15.5 = I_{\text{axis thro'}G} + 41\left(\frac{530}{1000}\right)^2$$

$$\therefore \quad 15.5 = I_{\text{axis thro'}G} + 11.5$$

$$\therefore \quad I_{\text{axis thro'}G} = 15.5 - 11.5$$

$$= 4 \text{ kg m}^2$$

Moments of inertia of various solids may be found in most engineering reference books, typical data being as shown below.

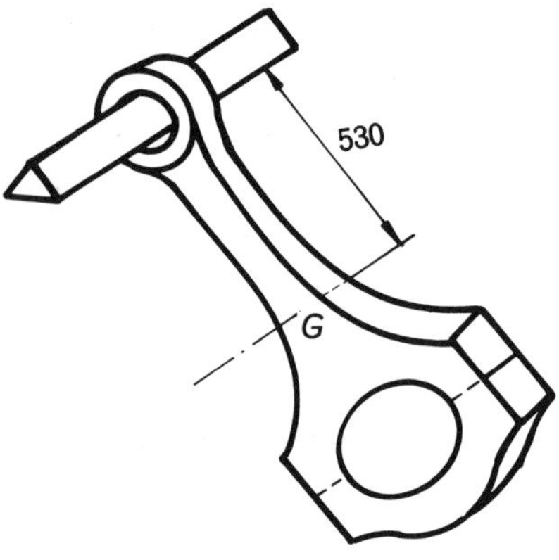

Fig. 12.6

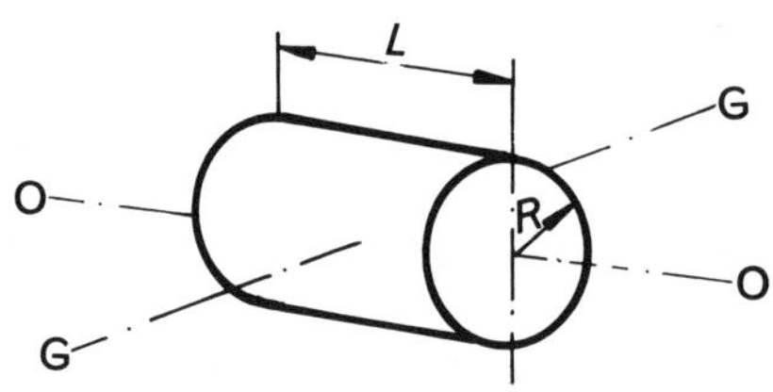

Fig. 12.7

Solid cylinder (Fig. 12.7)

$$\text{Polar } I_{OO} = \frac{1}{2}MR^2$$

$$I_{GG} = M\left(\frac{L^2}{12} + \frac{R^2}{4}\right)$$

where GG passes through the centre of gravity and is perpendicular to OO.

Solid rectangular block (Fig. 12.8)

$$I_{GG} = M\left(\frac{a^2+b^2}{12}\right)$$

where GG passes through the centre of gravity and is perpendicular to face *ab*.

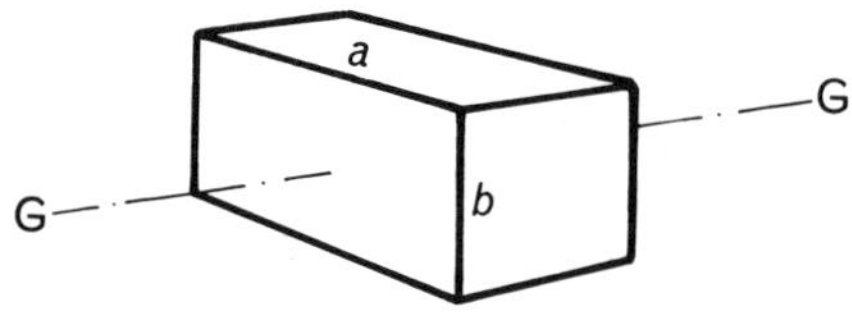

Fig. 12.8

Exercise 36

1) Find the polar moment of inertia of a solid aluminium cylinder 100 mm diameter and 500 mm long, if the density of aluminium is 2700 kg/m^3.

2) Find the radius of gyration k_{GG} of the solid rectangular block shown in Fig. 12.9 if axis GG is perpendicular to the face measuring 200 mm × 300 mm and passes through the centre of gravity of the block.

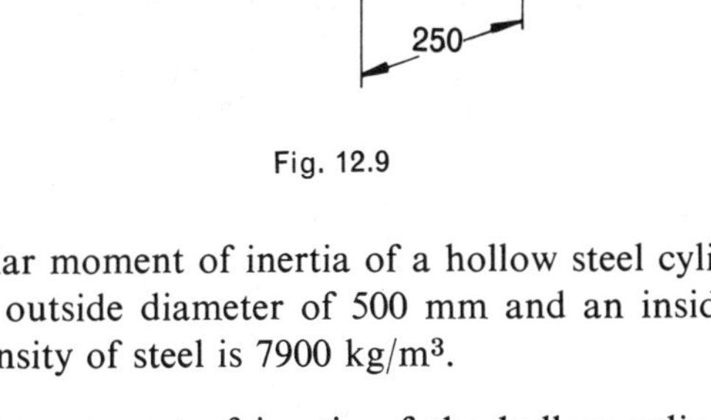

Fig. 12.9

3) Find the polar moment of inertia of a hollow steel cylinder, 750 mm long, having an outside diameter of 500 mm and an inside diameter of 400 mm. The density of steel is 7900 kg/m^3.

(*Hint:* The polar moment of inertia of the hollow cylinder is the difference between the polar moments of inertia of a solid cylinder 500 mm diameter and a solid cylinder 400 mm diameter.)

4) A hollow copper cylinder is shown in Fig. 12.10. If the density of copper is 9000 kg/m^3 find:

(a) I_{OO}

(b) I_{AA}

(*Hint:* when using the parallel axis theorem which gives $I_{AA} = I_{OO} + Mh^2$ remember that M is the mass of the hollow cylinder.)

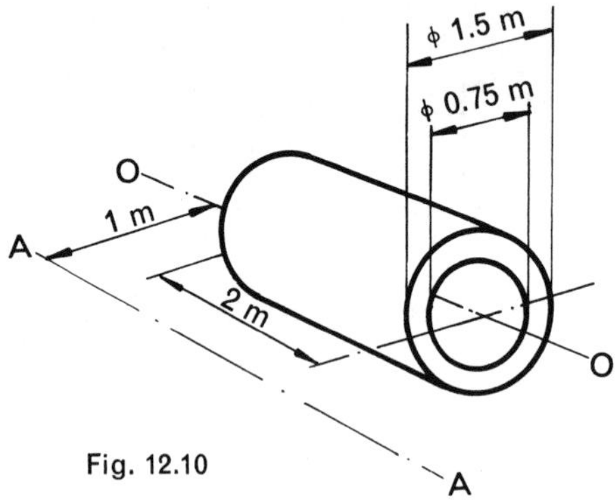

Fig. 12.10

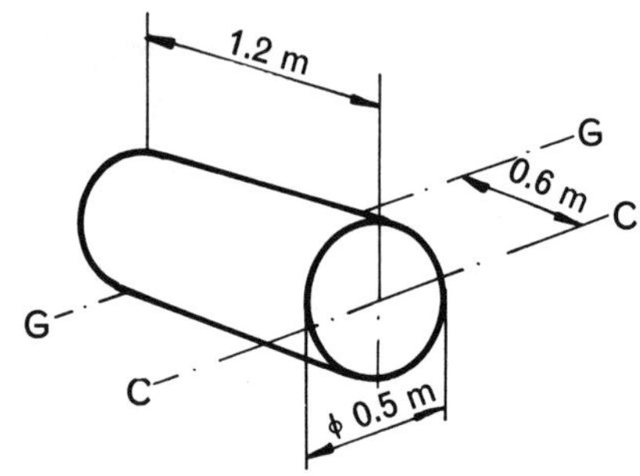

Fig. 12.11

5) Fig. 12.11 shows a solid copper cylinder which has a density of 9000 kg/m^3. Find:

(a) I_{GG}, where GG is an axis passing through the centre of gravity and perpendiculat to the polar axis.

(b) I_{CC}

6) Find I_{BB} for the solid block shown in Fig. 12.12 if the density of the material from which it is made is 8000 kg/m^3.

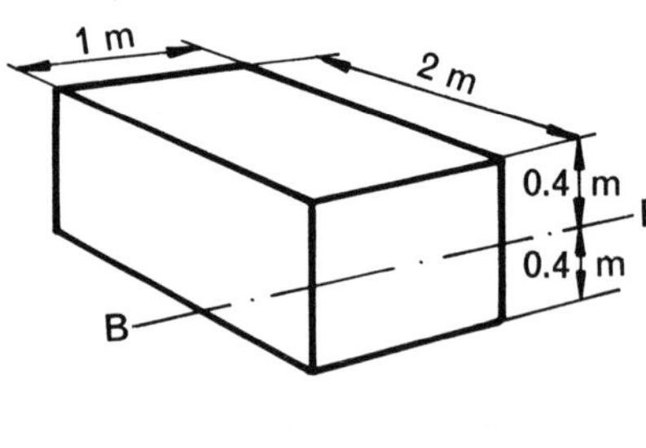

Fig. 12.12

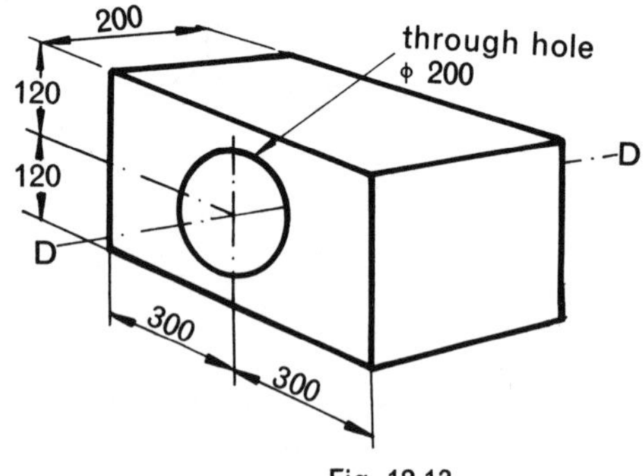

Fig. 12.13

7) The component shown in Fig. 12.13 is made from a material which has a density of 7500 kg/m^3. Find I_{DD}.

13. APPLICATIONS OF DIFFERENTIATION

After reaching the end of this chapter you should be able to:-

1. *Determine the second derivative* $\frac{d^2y}{dx^2}$, *or* $f''(x)$, *and higher derivatives of a function.*
2. *Evaluate a second derivative at a given point.*
3. *Calculate linear and angular velocity and acceleration at a given time using first and second derivatives.*
4. *Define a stationary point on a graph.*
5. *Determine the nature of a stationary point by considering the gradient on either side of the point.*
6. *Determine the nature of a stationary point using the sign of a second or higher derivative.*
7. *Solve problems involving maximum and minimum points.*

SECOND AND HIGHER DERIVATIVES

If $$y = x^6$$

then $$\frac{dy}{dx} = 6x^5$$

and if we differentiate this equation again with respect to x we obtain

$$\frac{d}{dx}\left(\frac{dy}{dx}\right) = \frac{d}{dx}(6x^5)$$

or $$\frac{d^2y}{dx^2} = 30x^4$$

Differentiating again gives

$$\frac{d}{dx}\left(\frac{d^2y}{dx^2}\right) = \frac{d}{dx}(30x^4)$$

or $$\frac{d^3y}{dx^3} = 120x^3$$

and for a fourth time

$$\frac{d^4y}{dx^4} = 360x^2$$

and so on.

Just as $\frac{dy}{dx} = 6x^5$ is called the first derived function or first differential coefficient or first derivative of y with respect to x, so $\frac{d^2y}{dx^2} = 30x^4$ is called the second derived function or second derivative of y with respect to x and so on. It should be noted that the 2 which occurs twice in $\dfrac{d^2y}{dx^2}$ is *not* an index but merely indicates that the original function has been differentiated that number of times.

Hence $\dfrac{d^2y}{dx^2}$ is *not* the same as $\left(\dfrac{dy}{dx}\right)^2$.

The optional function notation is: $y = f(x)$

$$\frac{dy}{dx} = f'(x)$$

$$\frac{d^2y}{dx^2} = f''(x)$$

$$\frac{d^3y}{dx^3} = f'''(x)$$

and so on.

EXAMPLE 1

Determine $\dfrac{dy}{dx}$ and $\dfrac{d^2y}{dx^2}$ when $y = x^3-2x^2+3x-7$.

Now $\qquad y = x^3-2x^2+3x-7$

$\therefore \qquad \dfrac{dy}{dx} = 3x^2-4x+3$

$\therefore \qquad \dfrac{d^2y}{dx^2} = 6x-4$

EXAMPLE 2

Evaluate $\dfrac{dy}{dx}$ and $\dfrac{d^2y}{dx^2}$ when $x = 2$ given that $y = \sqrt{x}+\sin x$.

Now $\qquad y = \sqrt{x}+\sin x$

$\qquad = x^{1/2}+\sin x$

$\therefore \qquad \dfrac{dy}{dx} = \tfrac{1}{2}x^{-1/2}+\cos x$

$\therefore \qquad \dfrac{d^2y}{dx^2} = -\tfrac{1}{4}x^{-3/2}-\sin x$

Before substituting the numerical value of x it is recommended that the expressions for $\dfrac{dy}{dx}$ and $\dfrac{d^2y}{dx^2}$ are transformed to give positive indices.

Although most electronic calculators will evaluate expressions with negative indices, mistakes may occur and it is more difficult to detect a computational error.

Remember also that sin 2 is the sine of 2 radians, *not* of 2 degrees, and that your calculator should be in radian mode.

Now $$\frac{dy}{dx} = \frac{1}{2x^{1/2}} + \cos x = \frac{1}{2\sqrt{x}} + \cos x$$

Hence when $x = 2$, $$\frac{dy}{dx} = \frac{1}{2\sqrt{2}} + \cos 2 = 0.3536 + (-0.4161)$$
$$= -0.0625$$

Also $$\frac{d^2y}{dx^2} = \frac{1}{4x^{3/2}} - \sin x = -\frac{1}{4(\sqrt{x})^3} - \sin x$$

and when $x = 2$, $$\frac{d^2y}{dx^2} = \frac{1}{4(\sqrt{2})^3} - \sin 2 = -0.0884 - 0.9093$$
$$= -0.9977$$

EXAMPLE 3

Given that $v = 4\sin\left(100\pi t + \frac{\pi}{3}\right)$ calculate the values of $\frac{dv}{dt}$ and $\frac{d^2v}{dt^2}$ when $t = 0.001$.

Now $$v = 4\sin\left(100\pi t + \frac{\pi}{3}\right)$$

Hence $$\frac{dv}{dt} = 4\cos\left(100\pi t + \frac{\pi}{3}\right) \times 100\pi \quad \text{(function of a function)}$$
$$= 400\pi\cos\left(100\pi t + \frac{\pi}{3}\right)$$

and $$\frac{d^2v}{d^2t} = -400\pi\sin\left(100\pi t + \frac{\pi}{3}\right) \times 100\pi$$
$$= -40\,000\pi^2\sin\left(100\pi t + \frac{\pi}{3}\right)$$

When $t = 0.001$: $$\frac{dv}{dt} = -40\,000\pi^2\sin(1.36) \quad \text{(angle in radians)}$$
$$\approx 263$$

and $$\frac{d^2v}{dt^2} = -40\,000\pi^2\sin(1.36)$$
$$\approx -386\,045$$

Exercise 37

1) If $y = 3x^3 + 2x - 7$ determine an expression for $\frac{d^2y}{dx^2}$ and evaluate it when $x = 3$.

2) Evaluate $f'(x)$ and $f''(x)$ when $x = -2$ given that $y = 5x^4 + 7x^2 + x$.

3) Given that $\theta = 18 + 20t^2 - 2t^3$ find the value of $\frac{d^2\theta}{dt^2}$ when $t = 5$.

4) If $i = -2\sin(20t - 0.02)$ evaluate $\frac{d^2i}{dt^2}$ when $t = 0.01$ (radian measure).

5) If $Z = 5\cos 3\theta$ find the value of $\frac{dZ}{d\theta}, \frac{d^2Z}{d\theta^2}$ and $\frac{d^3Z}{d\theta^3}$ when $\theta = \frac{\pi}{2}$.

6) Determine the value of $\frac{d^2p}{d\varnothing^2}$ given that $p = 6\sin 4\varnothing$, where $\varnothing$ is measured in radians and has a value equivalent to 28°.

7) If $f(\alpha) = 2\cos\frac{\alpha}{4}$ determine the value of $f''(\alpha)$ when $\alpha = 1.6$ radians.

8) Calculate the value of $\frac{d^2v}{dt^2}$ when $t = 0.002$ and

$$v = -5\sin\left(100\pi t - \frac{\pi}{6}\right).$$

9) When $t = \frac{\pi}{3}$ evaluate $f'(t)$ and $f''(t)$ given that $S = 10 + 5t - 10\sin t$.

10) Show that $\frac{dy}{dx} = \frac{d^2y}{dx^2}$ when $x = \frac{\pi}{8}$ and $y = \cos 2x - \sqrt{2}x$.

11) Show that the second derived function, $f''(\theta)$, of $f(\theta) = 2\theta^2 - \sin 2\theta$ is never decreasing.

VELOCITY AND ACCELERATION

Suppose that a vehicle starts from rest and travels 60 metres in 12 seconds. The average velocity may be found by dividing the total distance travelled by the total time taken, that is $\frac{60}{12} = 5$ m/s. This is NOT the INSTANTANEOUS velocity, however, AT a time of 12 seconds, but is the AVERAGE VELOCITY over the 12 seconds as calculated previously.

Fig. 13.1 shows a graph of distance s, against time t. The average velocity over a period is given by the gradient of the chord which meets the curve at the extremes of the period. Thus in the diagram the gradient of the dotted chord QR gives the average velocity between $t = 2$ s and $t = 6$ s. It is found to be $\frac{13}{4} = 3.25$ m/s.

The velocity at any point is the rate of change of s with respect t and may be found by finding the gradient of the curve at that point. In mathematical notation this is given by $\frac{ds}{dt}$.

Suppose we know that the relationship between s and t is

$$s = 0.417t^2$$

then velocity, $$v = \frac{ds}{dt} = 0.834t$$

and hence when $t = 12$ seconds, then $v = 0.834 \times 12 = 10$ m/s.

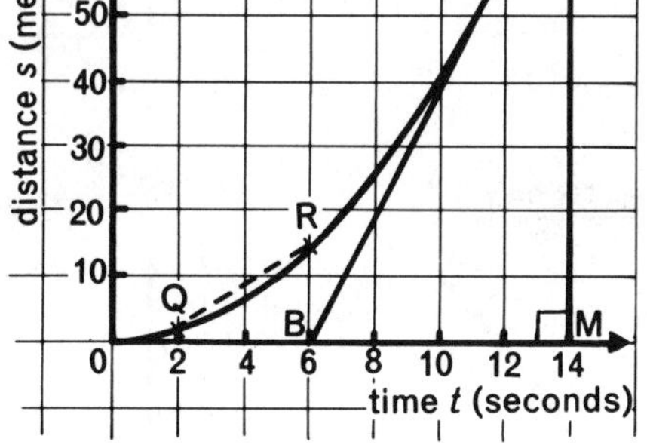

Fig. 13.1

This result may be found graphically by drawing the tangent to the curve of s against t at the point P and constructing a suitable right angled triangle ABM.

Hence the velocity at $P = \frac{AM}{BM} = \frac{80}{8} = 10$ m/s which verifies the theoretical result.

Similarly, the rate of change of velocity with respect to time is called acceleration and is given by the gradient of the velocity–time graph at any point. In mathematical notation this is given by $\frac{dv}{dt}$.

Now $$\frac{dv}{dt} = \frac{d}{dt}(v) = \frac{d}{dt}\left(\frac{ds}{dt}\right) = \frac{d^2s}{dt^2}$$

and so the acceleration, a, is given by either

$$\frac{dv}{dt} \text{ or } \frac{d^2s}{dt^2}.$$

The above reasoning was applied to linear motion, but it could also have been used for angular motion. The essential difference is that distance, s, is replaced by angle turned through, θ rad.

Both sets of results are summarised in Fig. 13.2.

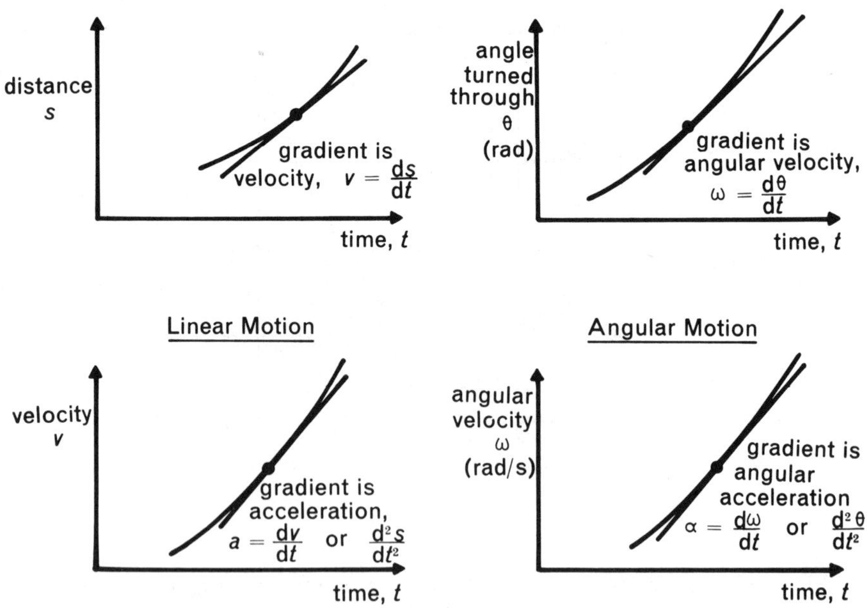

Fig. 13.2

EXAMPLE 4

A body moves a distance s metres in a time of t seconds so that $s = 2t^3-9t^2+12t+6$.

Find (a) its velocity after 3 seconds,
(b) its acceleration after 3 seconds, and
(c) when the velocity is zero.

We have, $$s = 2t^3-9t^2+12t+6$$

$\therefore$ $$\frac{ds}{dt} = 6t^2-18t+12$$

and $$\frac{d^2s}{dt^2} = 12t-18$$

(a) When $t = 3$ then velocity is $\dfrac{ds}{dt} = 6(3)^2-18(3)+12 = 12$ m/s

(b) when $t = 3$ then acceleration is $\dfrac{d^2s}{dt^2} = 12(3)-18 = 18$ m/s^2

(c) when the velocity is zero then $\dfrac{ds}{dt} = 0$

that is $$6t^2-18t+12 = 0$$

$\therefore$ $$t^2-3t+2 = 0$$

$\therefore$ $$(t-1)(t-2) = 0$$

$\therefore$ either $t-1 = 0$ or $t-2 = 0$

$\therefore$ either $t = 1$ second or $t = 2$ seconds.

EXAMPLE 5

The angle θ radians is connected with the time t seconds by the relationship $\theta = 20+5t^2-t^3$.

Find (a) the angular velocity when $t = 2$ seconds, and
(b) the value of t when the angular deceleration is 4 rad/s^2.

We have, $\theta = 20+5t^2-t^3$

$\therefore$ $\dfrac{d\theta}{dt} = 10t-3t^2$

and $\dfrac{d^2\theta}{dt^2} = 10-6t$

(a) when $t = 2$, then the angular velocity $\dfrac{d\theta}{dt} = 10(2)-3(2)^2 = 8$ rad/s.

(b) an angular deceleration of 4 rad/s^2 may be called an angular acceleration of -4 rad/s^2,

$\therefore$ when $\dfrac{d^2\theta}{dt^2} = -4$ then $-4 = 10-6t$

or $t = 2.33$ seconds.

Exercise 38

1) If $s = 10+50t-2t^2$, where s metres is the distance travelled in t seconds by a body, what is the velocity of the body after 2 seconds?

2) If $v = 5+24t-3t^2$ where v m/s is the velocity of a body at a time t seconds, what is the acceleration when $t = 3$?

3) A body moves s metres in t seconds where $s = t^3-3t^2-3t+8$.

Find (a) its velocity at the end of 3 seconds,

(b) when its velocity is zero,

(c) its acceleration at the end of 2 seconds,

(d) when its acceleration is zero.

4) A body moves s metres in t seconds where $s = \dfrac{1}{t^2}$. Find the velocity and acceleration after 3 seconds.

5) The distance s metres travelled by a falling body starting from rest after a time t seconds is given by $s = 5t^2$. Find its velocity after 1 second and after 3 seconds.

6) The distance s metres moved by the end of a lever after a time t seconds is given by the formula $s = 6t^2$. Find the velocity of the end of the lever when it has moved a distance $\frac{1}{2}$ metre.

7) The angular displacement θ radians of the spoke of a wheel is given by the expression $\theta = \dfrac{1}{2}t^4-t^3$ where t seconds is the time.

Find (a) the angular velocity after 2 seconds,

(b) the angular acceleration after 3 seconds,

(c) when the angular acceleration is zero.

8) An angular displacement θ radians in time t seconds is given by the equation $\theta = \sin 3t$.

Find (a) the angular velocity when $t = 1$ second,

(b) the smallest positive value of t for which the angular velocity is 2 rad/s,

(c) the angular acceleration when $t = 0.5$ seconds,

(d) the smallest positive value of t for which the angular acceleration is 9 rad/s^2.

9) A mass of 5000 kg moves along a straight line so that the distance s metres travelled in a time t seconds is given by $s = 3t^2+2t+3$. If v m/s is its velocity and m kg is its mass then its kinetic energy is given by the formula $\frac{1}{2}mv^2$. Find its kinetic energy at a time $t = 0.5$ seconds remembering that the joule (J) is the unit of energy.

CRITICAL POINTS

Any point on the graph of $y = f(x)$ such that $f'(x) = 0$, or $\frac{dy}{dx} = 0$, is called a **critical point** on the curve. At this point the tangent is parallel to the x-axis and for this reason the term **stationary point** is often used instead of critical point.

The nature of the curve in the neighbourhood of a critical point is important.

It is possible at this point to have a **local maximum turning point**, or a **local minimum turning point** or a **local point of horizontal inflexion**.

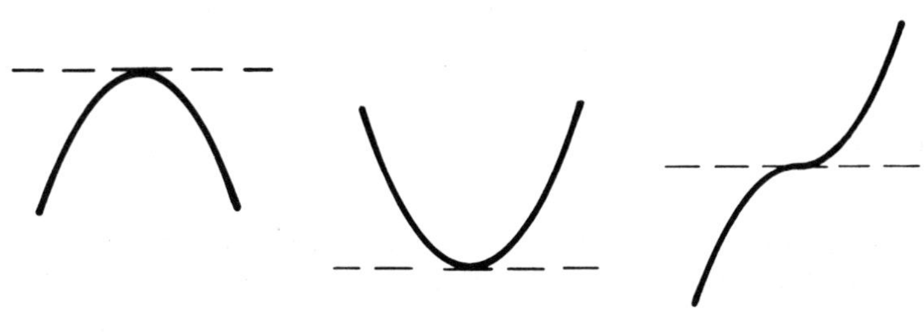

Maximum and minimum turning points and problems will be investigated first and then the theory will be expanded to include points of inflexion and curve sketching.

TURNING POINTS

At the points P and Q (Fig. 13.3) the tangent to the curve is parallel to the x-axis. The points P and Q are called *turning points*. The turning point at P is called a *maximum* turning point and the turning point at Q is called a *minimum* turning point. It will be seen from Fig. 13.3 that the value of y at P is not the greatest value of y nor is the value of y at Q the least. The terms maximum and minimum values apply only to the values of y at the turning points and not to the values of y in general.

In practical applications, however, we are usually concerned with a specific range of values of x which are dictated by the problem. There is then no difficulty in identifying a particular maximum or minimum within this range of values of x.

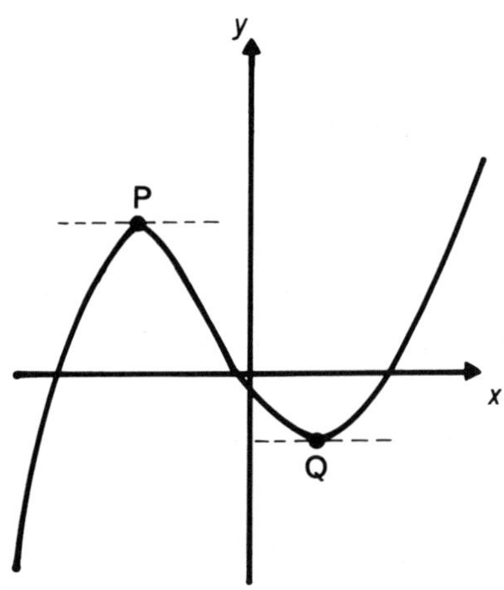

Fig. 13.3

EXAMPLE 6

Plot the graph of $y = x^3 - 5x^2 + 2x + 8$ for values of x between -2 and 6. Hence find the maximum and minimum values of y.

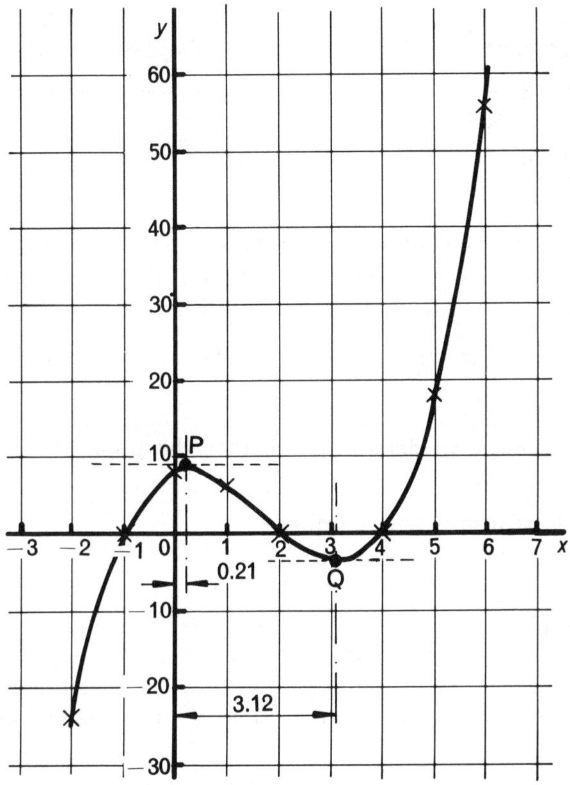

Fig. 13.4

To plot the graph we draw up a table in the usual way.

x	−2	−1	0	1	2	3	4	5	6
$y = x^3 - 5x^2 + 2x + 8$	−24	0	8	6	0	−4	0	18	56

The graph is shown in Fig. 13.4. The maximum value occurs at the point P where the tangent to the curve is parallel to the x-axis. The minimum value occurs at the point Q where again the tangent to the curve is parallel to the x-axis. From the graph the maximum value of y is 8.21 and the minimum value of y is −4.06.

Notice that the value of y at P is not the greatest value of y nor is the value of y at Q the least. However, the values of y at P and Q are called the maximum and minimum values of y respectively.

Fig. 13.5

It is not always convenient to draw the full graph to find the turning points as in the previous example. At a turning point the tangent to the curve is parallel to the x-axis (Fig. 13.5) and hence the gradient of the curve is zero, i.e. $\frac{dy}{dx} = 0$. Using this fact enables us to find the values of x at which the turning points occur.

It is then necessary to determine whether the points are maximum or minimum.

Two methods of testing are as follows:

Method 1

Consider the gradients of the curve on either side of the turning point. Fig. 13.6 shows how the gradient (or slope) of curve changes in the vicinity of a turning point.

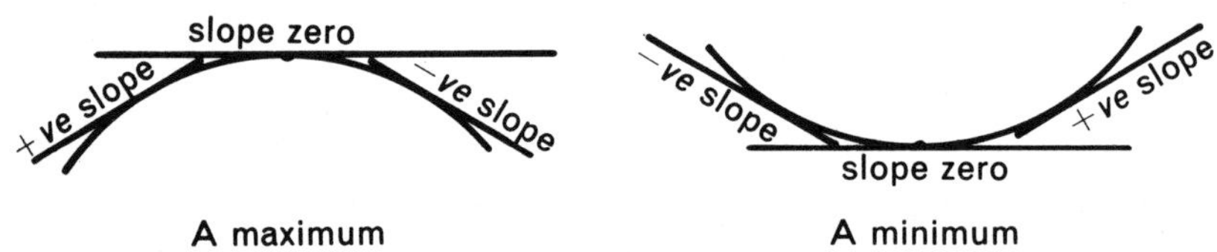

Fig. 13.6

Method 2

Find the value of $\frac{d^2y}{dx^2}$ at the turning point.

If it is positive then the turning point is a minimum, and if it is negative then the turning point is a maximum.

If the original expression may be differentiated twice and the expression for $\frac{d^2y}{dx^2}$ obtained without too much difficulty, then the second method is generally used.

EXAMPLE 7

Find the maximum and minimum values of y given that $y = x^3+3x^2-9x+6$.

We have $\quad y = x^3+3x^2-9x+6$

$\because \quad \dfrac{dy}{dx} = 3x^2+6x-9$

and $\quad \dfrac{d^2y}{dx^2} = 6x+6$

At a turning point $\quad \dfrac{dy}{dx} = 0$

$\therefore \quad 3x^2+6x-9 = 0$

$\therefore \quad x^2+2x-3 = 0$ by dividing through by 3.

$\therefore \quad (x-1)(x+3) = 0$

$\therefore \quad$ either $x-1 = 0$ or $x+3 = 0$

$\therefore \quad$ either $x = 1$ or $x = -3$

Test for maximum or minimum:

From above we have $\quad \dfrac{d^2y}{dx^2} = 6x+6$

$\therefore \quad$ at the point where $x = 1$, $\dfrac{d^2y}{dx^2} = 6(1)+6 = +12$

This is positive and hence the turning point at $x = 1$ is a minimum.

The minimum value of y may be found by substituting $x = 1$ into the given equation. Hence:

$$y_{min} = (1)^3+3(1)^2-9(1)+6 = 1.$$

At the point where $x = -3$, $\dfrac{d^2y}{dx^2} = 6(-3)+6 = -12.$

This is negative and hence at $x = -3$ there is a maximum turning point. The maximum value of y may be found by substituting $x = -3$ into the given equation. Hence:

$$y_{max} = (-3)^3+3(-3)^2-9(-3)+6 = +33.$$

To illustrate the test for maximum and minimum using the tangent gradient method, this method will be used to verify the above results.

At the turning point where $x = 1$, we know that

$$\frac{dy}{dx} = 0, \quad \text{i.e. there is zero slope,}$$

and using a value of x slightly less than 1, say $x = 0.5$ then

$$\frac{dy}{dx} = 3(0.5)^2+6(0.5)-9 = -5.25 \quad \text{i.e. there is a negative slope,}$$

and using a value of x slightly greater than 1, say $x = 1.5$ then

$$\frac{dy}{dx} = 3(1.5)^2+6(1.5)-9 = +6.75 \quad \text{i.e. there is a positive slope.}$$

These results are best shown by means of a diagram (Fig. 13.7) which indicates clearly that when $x = 1$ we have a minimum.

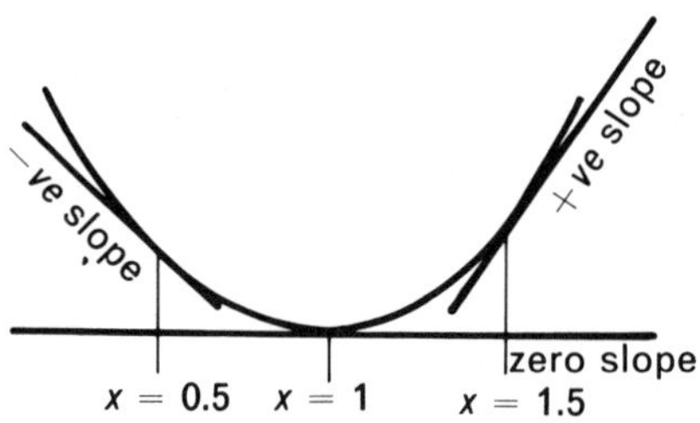

Fig. 13.7

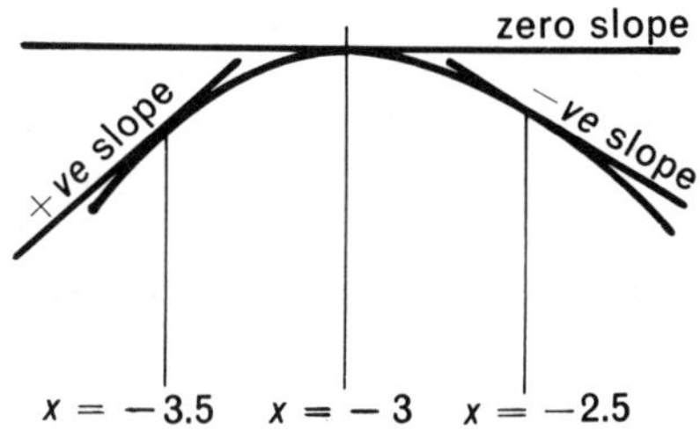

Fig. 13.8

Now at the turning point where $x = -3$ we know that

$$\frac{dy}{dx} = 0, \quad \text{i.e. there is zero slope,}$$

and using a value of x slightly less than -3, say $x = -3.5$ then

$$\frac{dy}{dx} = 3(-3.5)^2+6(-3.5)-9 = +6.75 \quad \text{i.e. there is a positive slope,}$$

and using a value of x slightly greater than -3, say $x = -2.5$ then

$$\frac{dy}{dx} = 3(-2.5)^2+6(-2.5)-9 = -5.25 \quad \text{i.e. there is a negative slope.}$$

Fig. 13.8 indicates that when $x = -3$ we have a maximum turning point.

There are many applications in engineering which involve the finding of maxima and minima. The first step is to construct an equation connecting the quantity for which a maximum or minimum is required in terms of another variable. A diagram representing the problem may help in the formation of this initial equation.

EXAMPLE 8

A rectangular sheet of metal 360 mm by 240 mm has four equal squares cut out at the corners. The sides are then turned up to form a rectangular box. Find the length of the sides of the squares cut out so that the volume of the box may be as great as possible, and find this maximum volume.

Let the length of the side of each cut away square be x mm as shown in Fig. 13.9.

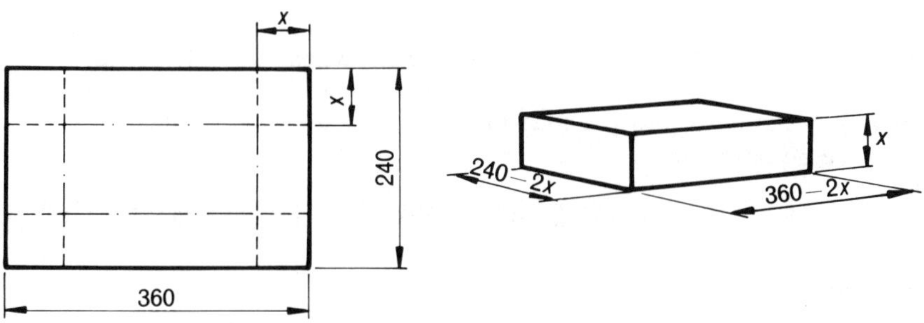

Fig. 13.9

Hence the volume is $\qquad V = x(240-2x)(360-2x)$

$$= 4x^3-1200x^2+86\,400x$$

$$\therefore \qquad \frac{dV}{dx} = 12x^2-2400x+86\,400$$

and $\qquad \dfrac{d^2V}{dx^2} = 24x-2400$

At a turning point $\qquad \dfrac{dV}{dx} = 0$

$$\therefore \qquad 12x^2-2400x+86\,400 = 0$$

or $\qquad x^2-200x+7200 = 0$ by dividing through by 12.

Now this is a quadratic equation which does not factorise so we will have to solve using the formula for the standard quadratic $ax^2+bx+c = 0$ which gives $x = \dfrac{-b\pm\sqrt{b^2-4ac}}{2a}$. Hence the solution of our equation is $$x = \frac{-(-200)\pm\sqrt{(-200)^2-4\times1\times7200}}{2\times1}$$

$\therefore$ either $x = 152.9$ or $x = 47.1$

However, from the physical sizes of the sheet, it is not possible for x to be 152.9 mm (since one side is only 240 mm long) so we reject this solution. Hence $x = 47.1$ mm.

Test for maximum or minimum:

From the above we have $$\frac{d^2V}{dx^2} = 24x-2400$$

and hence when $x = 47.1$, $$\frac{d^2V}{dx^2} = 24(47.1)-2400 = -1270$$

This is negative and hence V is a maximum when $x = 47.1$ mm.

It only remains to find the maximum volume by substituting $x = 47.1$ into the equation for V.

$$\therefore \quad V_{max} = 47.1(240-2\times47.1)(360-2\times47.1) = 1.825\times10^6 \text{ mm}^3.$$

EXAMPLE 9

A cylinder with an open top has a capacity of 2 m³ and is made from sheet metal. Neglecting any overlaps at the joints find the dimensions of the cylinder so that the amount of sheet steel used is a minimum.

Let the height of the cylinder be h metres and the radius of the base be r metres as shown in Fig. 13.10.

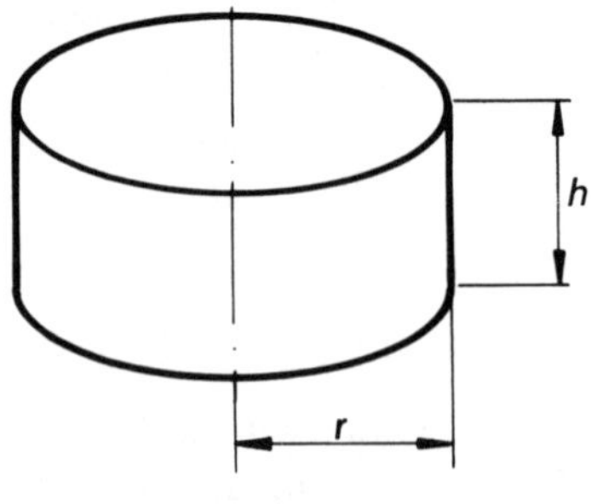

Fig. 13.10

Now the total area of metal = area of base+area of curved side:

$\therefore \quad A = \pi r^2 + 2\pi rh$

We cannot proceed to differentiate as there are two variables on the right hand side of the equation. It is possible, however, to find a connection between r and h using the fact that the volume is 2 m².

Now $\quad$ volume of a cylinder $= \pi r^2 h$

$\therefore \quad 2 = \pi r^2 h$

from which $\quad h = \dfrac{2}{\pi r^2}$

We may now substitute for h is the equation for A,

$$\therefore \quad A = \pi r^2 + 2\pi r\left(\frac{2}{\pi r^2}\right)$$

$$\therefore \quad = \pi r^2 + \frac{4}{r}$$

$$\therefore \quad = \pi r^2 + 4r^{-1}$$

$$\therefore \quad \frac{dA}{dr} = 2\pi r - 4r^{-2}$$

and $\quad \dfrac{d^2A}{dr^2} = 2\pi + 8r^{-3}$

Now for a turning point $\quad \dfrac{dA}{dr} = 0$

or $\quad 2\pi r - 4r^{-2} = 0$

$$\therefore \quad 2\pi r - \frac{4}{r^2} = 0$$

$$\therefore \quad 2\pi r = \frac{4}{r^2}$$

$$\therefore \quad r^3 = \frac{2}{\pi} = 0.637$$

$$r = \sqrt[3]{0.637} = 0.860$$

To test for a minimum:

From above we have $\quad \dfrac{d^2A}{dr^2} = 2\pi + 8r^{-3}$

$$= 2\pi + \frac{8}{r^3}$$

We do not need to do any further calculation here as this expression must be positive for all positive values of r. Hence $r = 0.86$ makes A a minimum.

We may find the corresponding value of h by substituting $r = 0.86$ into the equation found previously for h in terms of r

$$\therefore \qquad h = \frac{2}{\pi(0.86)^2} = 0.86$$

hence for the minimum amount of metal to be used the radius is 0.86 m and the height is 0.86 m.

Exercise 39

1) Find the maximum and minimum values of:

(a) $y = 2x^3-3x^2-12x+4$

(b) $y = x^3-3x^2+4$

(c) $y = 6x^2+x^3$

2) Given that $y = 60x+3x^2-4x^3$, calculate:

(a) the gradient of the tangent to the curve of y at the point where $x = 1$;

(b) the value of x for which y has its maximum value;

(c) the value of x for which y has its minimum value.

3) Calculate the co-ordinates of the points on the curve $y = x^3-3x^2-9x+12$ at each of which the tangent to the curve is parallel to the x-axis.

4) A curve has the equation $y = 8+2x-x^2$. Find:

(a) the value of x for which the gradient of the curve is 6;

(b) the value of x which gives the maximum value of y;

(c) the maximum value of y.

5) The curve $y = 2x^2+\dfrac{k}{x}$ has a gradient of 5 when $x = 2$. Calculate (a) the value of k; (b) the minimum value of y.

6) From a rectangular sheet of metal measuring 120 mm by 75 mm equal squares of side x are cut from each of the corners. The remaining flaps are then folded upwards to form an open box. Prove that the volume of the box is given by $V = 9000x-390x^2+4x^3$. Find the value of x such that the volume is a maximum.

7) An open rectangular tank of height h metres with a square base of side x metres is to be constructed so that it has a capacity of 500 cubic metres. Prove that the surface area of the four walls and the base will be $\left(\frac{2000}{x}+x^2\right)$ square metres. Find the value of x for this expression to be a minimum.

8) The volume of a cone is given by the formula $V = \frac{1}{3}\pi r^2 h$, where h is the height of the cone and r its radius. If $h = 6-r$, calculate the value of r for which the volume is a maximum.

9) A box without a lid has a square base of side x mm and rectangular sides of height h mm. It is made from 10 800 mm^2 of sheet metal of negligible thickness. Prove that $h = \frac{10\,800-x^2}{4x}$ and that the volume of the box is $(2700x-\frac{1}{4}x^3)$. Hence calculate the maximum volume of the box.

10) A cylindrical tank, with an open top, is to be made to hold 300 cubic metres of liquid. Find the dimensions of the tank so that its surface area shall be a minimum.

11) A cooling tank is to be made with the trapezoidal section as shown:

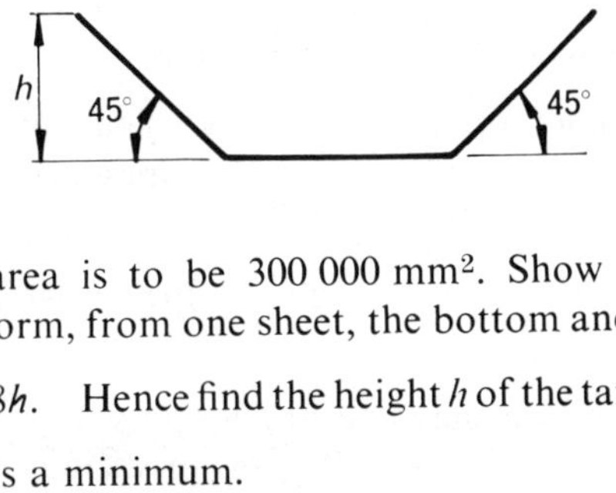

Its cross-sectional area is to be 300 000 mm^2. Show that the width of material needed to form, from one sheet, the bottom and folded-up sides is $w = \frac{300\,000}{h}+1.828h$. Hence find the height h of the tank so that the width of material needed is a minimum.

12) A cylindrical cup is to be drawn from a disc of metal of 50 mm diameter. Assuming that the surface area of the cup is the same as that of the disc find the dimensions of the cup so that its volume is a maximum.

13) A lever weighing 12 N per m run of its length is as shown:

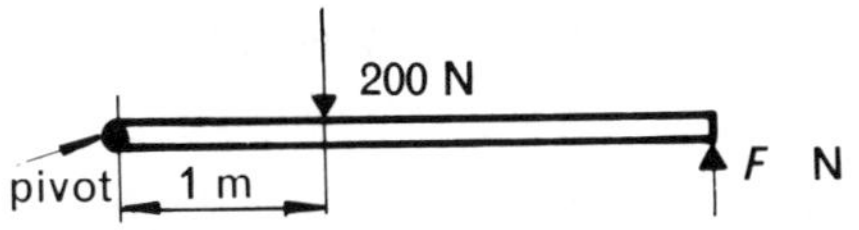

Find the length of the lever so that the force F shall be a minimum.

14. BOOLEAN ALGEBRA AND COMPUTING

After reaching the end of this chapter you should be able to:-

1. *Use the laws of set theory when applied to the operators '+' and '.'*
2. *Use the laws of set theory to simplify sets.*
3. *Demonstrate the equality of sets using Venn diagrams.*
4. *Identify the symbols used for binary logic elements (gates).*
5. *Use logic gates in electronics and hydraulics.*
6. *Write arithmetic expressions in BASIC.*
7. *Use the conditional IF . . . THEN stateme in programming.*
8. *Draw flowcharts.*
9. *Write programmes using the FOR . . . NE statement.*
10. *Use the DIMension statement in programm*

INTRODUCTION

Since 1940 Boolean algebra has played an increasingly significant role in the design of switching circuits and other two-state devices. The two states are usually represented by the symbols 0 and 1.

In computers and microprocessors binary signals are used. These signals can only have two levels whose actual values depend upon the circuit components.

For instance, a transistor circuit might use the levels -3 volts (to represent 0) and $+3$ (to represent 1).

The various circuits in a machine like a computer use logic elements or gates such as NOT, AND and OR gates in various combinations. These gates combine and manipulate the binary signals so that the required functions are performed.

SET THEORY

Since every Boolean algebra may be interpreted as an algebra of sets, for some choice of universal set, the following notation and laws must be known.

1) The elements of a set are denoted by $a, b, c, \ldots$ and sets by $A, B, C, \ldots$

2) When listing the elements of a set we use { } notation.

3) Elements *not* in set A are denoted by $\bar{A}$. Thus NOT $A = \bar{A}$.

4) If the set A is contained in the set B we write $A \subset B$. If at the same time $B \subset A$ then $A = B$.

5) The set containing *all* the elements under discussion is called the **universal** set and is denoted by E (entirety), U (universal) or by the symbol 1. In this book the symbol 1 is used and it should not be confused with the number 1.

6) The set containing *no* elements is called the **null** set and is denoted by the symbols { }, ø or 0. In this book the symbol 0 is used and it should not be confused with the number 0.

7) A set containing 3 elements has 2^3 subsets; that is 8 subsets. A set containing 4 elements has 2^4 subsets; and so on.

COMBINATION OF SETS

Union

The *union* of two sets is denoted by $+$ and we define

$$A+B = \{x: x \in A \quad \text{or} \quad x \in B \quad \text{or} \quad x \in \text{both sets.}\}$$

EXAMPLE 1

$A = \{0,1,2,3\}$ and $B = \{3,4,5\}$

Hence $A+B = \{0,1,2,3,4,5\}$

Intersection

The *intersection* of two sets is denoted by . and we define

$$A.B = \{x: x \in A \quad \text{and} \quad x \in B\}$$

EXAMPLE 2

$$A = \{0,1,2,3\} \quad \text{and} \quad B = \{3,4,5\}$$

Hence $A.B = \{3\}$

LAWS OF SET THEORY APPLIED TO THE OPERATORS '+' AND '.'

The following laws are very helpful and the results should be memorised. Each law can be verified using a Venn diagram.

Law	+	.
1. Tautology	$A+A = A$	$A.A = A$
2. Absorption	$A+A.B = A$	$A.(A+B) = A$
3. Complementation	$A+\overline{A} = 1$	$A.\overline{A} = 0 \quad \overline{\overline{A}} = A$
4. De Morgan	$\overline{(A+B)} = \overline{A}.\overline{B}$	$\overline{A.B} = \overline{A}+\overline{B}$
5. Commutative	$A+B = B+A$	$A.B = B.A$
6. Associative	$A+(B+C) = (A+B)+C$	$A.(B.C) = (A.B).C$
7. Distributive	$A.(B+C) = A.B+A.C$	$A+B.C = (A+B).(A+C)$

Operations with 0 and 1

(a) $1+A = 1$ (b) $0.A = 0$ (c) $0+A = A$

(d) $1.A = A$ (e) $\overline{1} = 0$ (f) $\overline{0} = 1$

Duality

Examination of the above laws show that in any identity in the algebra of sets if we replace each . by + and each + by ., each 0 by 1 and each 1 by 0, then the resulting expression is also true.

EXAMPLE 3

The distributive law states $A.(B+C) = A.B+A.C$

Replacing each ., +, 0 and 1 as explained above gives

$$A+(B.C) = (A+B).(A+C)$$

which is the other distributive law.

SIMPLIFICATION OF SETS

The methods usually employed are:

1) Expansion and simplification.

2) Factorisation.

EXAMPLE 4

(a) Expand and simplify the expression $X+\overline{X}.Y$. Use a Venn diagram to verify the result.

$$X+\overline{X}.Y=(X+\overline{X}).(X+Y) \quad \ldots \text{ Distributive law}$$
$$= \quad 1 \quad .(X+Y)$$
$$= X+Y$$

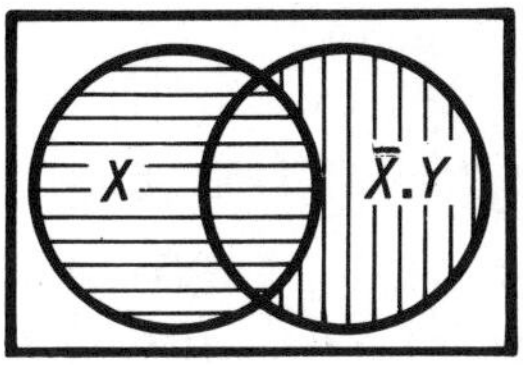

$X+\overline{X}.Y$

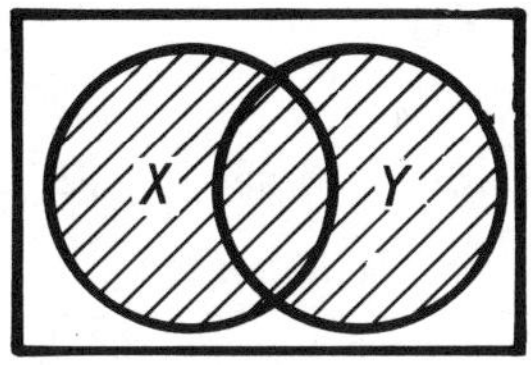

$X+Y$

(b) Simplify the expression $A(\overline{A}+B)+B(B+C)+B$.

$$A(\overline{A}+B)+B(B+C)+B = A\overline{A}+AB+BB+BC+B$$
$$= 0+AB+B+BC+B$$
$$= B+AB+BC$$
$$= B(1+A+C)$$
$$= B.1$$
$$= B$$

With practice many of the intermediate steps can be eliminated. Note that AB should be written $A.B$ and terms like $B(B+C)$ should be written $B.(B+C)$. We assume the . to be in position.

(c) Simplify $\overline{XY+\overline{X}Y+X\overline{Y}}$ to a single term and verify the result by drawing a Venn diagram.

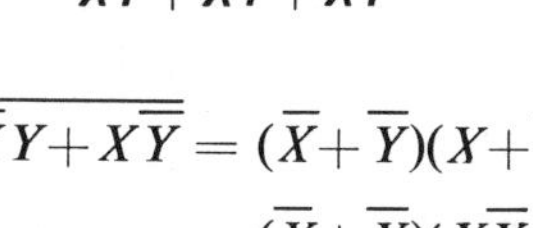

$\overline{XY+\overline{X}Y+X\overline{Y}}$

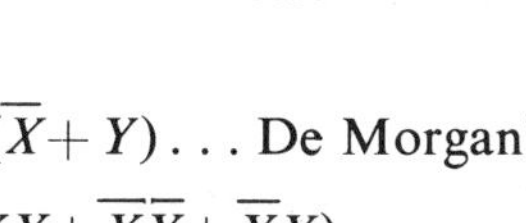

$\overline{X}.\overline{Y}$

$$\overline{XY+\overline{X}Y+X\overline{Y}} = (\overline{X}+\overline{Y})(X+\overline{Y})(\overline{X}+Y) \ldots \text{ De Morgan's law}$$
$$= (\overline{X}+\overline{Y})(X\overline{X}+XY+\overline{X}\overline{Y}+\overline{Y}Y)$$
$$= \{\overline{X}+\overline{Y})(XY+\overline{X}\overline{Y})$$
$$= \overline{X}XY+\overline{X}\overline{Y}+XY\overline{Y}+\overline{X}\overline{Y}$$
$$= \overline{X}\overline{Y}$$

Using cross-hatching or colour pencils we see that both expressions represent the same area.

Exercise 40

1) Given that $A = \{3,5,7,9,11\}$, $B = \{3,4,5,6,7,8\}$, $C = \{2,4,6,8,10\}$ and that the universal set is $\{2,3,4,5,6,7,8,9,10,11\}$, find:

(a) $A+B$ (b) $B.C$ (c) $\overline{A}.C$ (d) $A.(B+C)$
(e) $\overline{A.(B+C)}$ (f) $\overline{A}+\overline{B}.\overline{C}$ (g) $\overline{A}.\overline{B}$ (h) $\overline{A.B}$

2) Use the laws of set theory to simplify each of the following expressions to a single term:

(a) $A.\overline{B}.\overline{A}.\overline{B}.$ (b) $AB+A\overline{B}+\overline{A}B+\overline{A}\overline{B}$ (c) $A\overline{C}+ABC+AC$
(d) $(X+Y)(\overline{X}+Y)$ (e) $(X+XY+XYZ)(X+Y+Z)$

3) Draw Venn diagrams to show the equivalence of the following results:

(a) $\overline{X.Y} = \overline{X}+\overline{Y}$ (b) $\overline{X}.\overline{Y} = \overline{X+Y}$
(c) $(X\overline{Y}+YZ)(\overline{X}\overline{Y}+XZ+YZ) = XZ+YZ$

4) Expand and simplify the following set expressions:

(a) $(A+B)(\overline{A}+B)(A+\overline{B})(\overline{A}+\overline{B})$
(b) $X+YW+(W+Y)X+(Z+W)(X+Y)$

LOGIC ELEMENTS

A circuit which consists of two switches, A and B, connected in parallel will be noted by $A+B(A$ OR $B)$, and a circuit consisting of two switches, A and B, connected in series will be denoted by $A.B$ (A AND B). If one switch is always open when another is closed we will denote one by A and the other by $\overline{A}$, (NOT A).

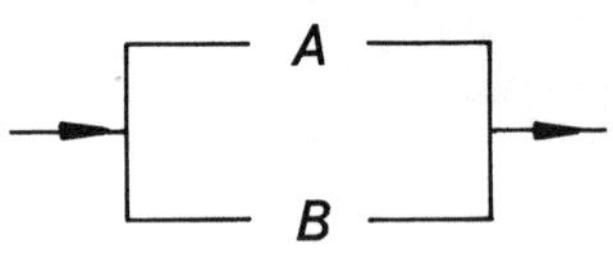

$A+B$

$A.B$

Thus to each series–parallel circuit there corresponds an algebraic expression and, conversely, to each algebraic expression, involving only + and . , there corresponds a series–parallel circuit.

A closed switch is represented by the symbol 1 and an open switch by the symbol 0. Various types of binary logic elements or gates are used in these transistorised circuits, the most common being the NOT (NEGATER), OR, NOR, AND and NAND gates.

Gate	Symbol	
NOT (NEGATER)	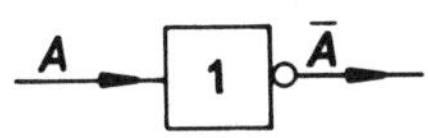	When the input signal represents 1 the output signal represents 0, and vice versa.
OR	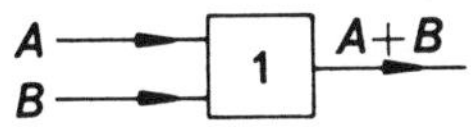	The output signal is represented by 1 only when one or more of the input signals is represented by 1.
NOR	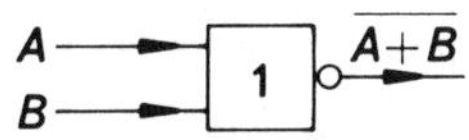	Output $\overline{A+B} = \overline{A}.\overline{B}$ (De Morgan's law)
AND	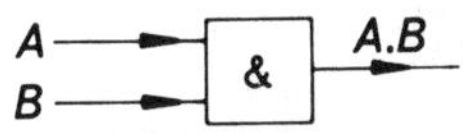	The output signal is represented by 1 only when ALL input signals are represented by 1.
NAND	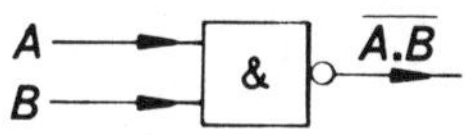	Output $\overline{A.B} = \overline{A}+\overline{B}$ (De Morgan's law)

EXAMPLE 5

Draw a logic diagram to represent the Boolean function $f = A+BC$. We require an AND element for the term $B.C$ (B AND C), and an OR element to combine both terms.

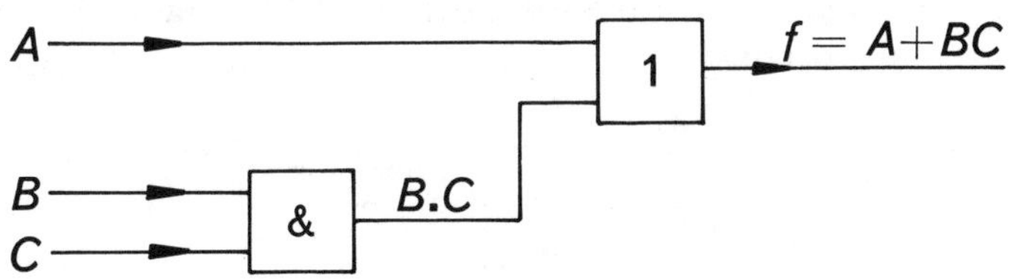

EXAMPLE 6

A particular output can usually be achieved in a variety of ways depending on the availability or cost of individual gates. Use the laws of the algebra of sets to obtain three different outputs each equivalent to the function $f = \overline{A}.\overline{B}+\overline{A}.\overline{C}$ and hence draw the three logic diagrams.

Using De Morgan's law twice we obtain the three different functions each with an equivalent output.

$$f = \overline{A}.\overline{B} + \overline{A}.\overline{C} \qquad \text{Fig. 14.1 (a)}$$

$$f = \overline{(A+B)} + \overline{(A+C)} \qquad \text{Fig. 14.1 (b)}$$

$$f = \overline{(A+B)(A+C)} \qquad \text{Fig. 14.1 (c)}$$

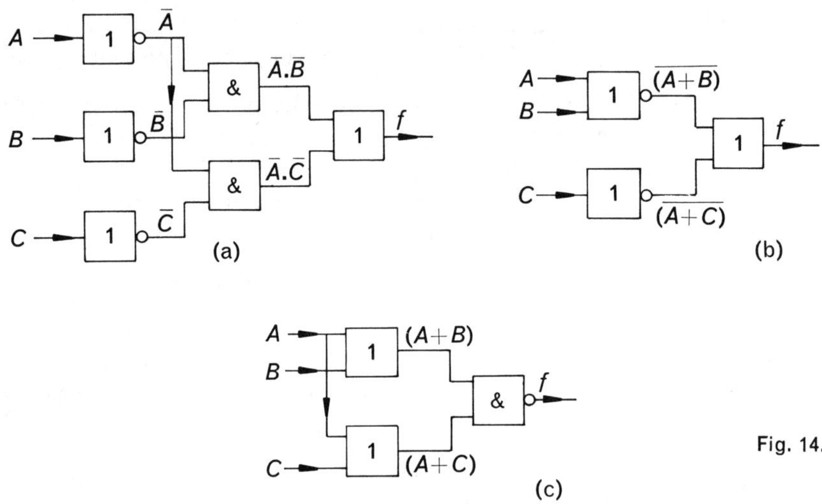

Fig. 14.1

FLUID LOGIC

The above theory can be applied to fluid networks. For example Fig. 14.2 shows a block diagram of a three-way pneumatic valve from which, depending on the position of the valve, we either have an outlet denoted by 1, or no outlet denoted by 0.

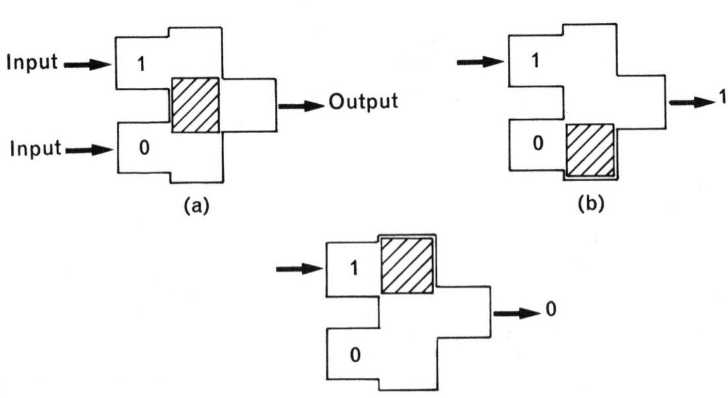

Fig. 14.2

Fig. 14.2(b) and (c) show an example of a NOT gate since changing the position of the valve will change the flow conditions.

Figs. 14.3(a), (b) and (c) show how OR, AND and NOR gates can be achieved using valves of this type.

(a)

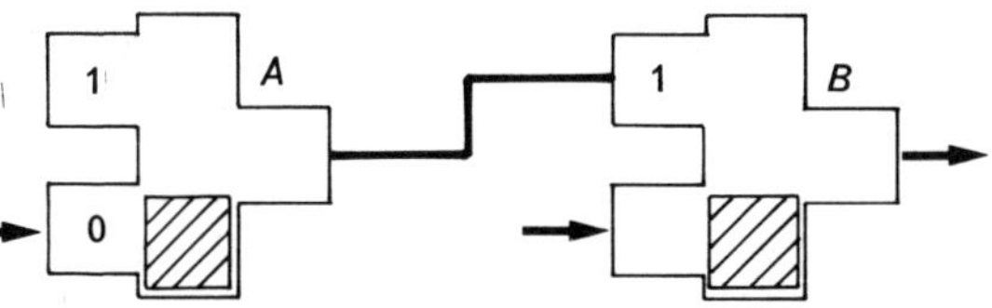

A OR $B = A+B$.
The application of a pressure signal to either valve will cause an output to occur.

(b)

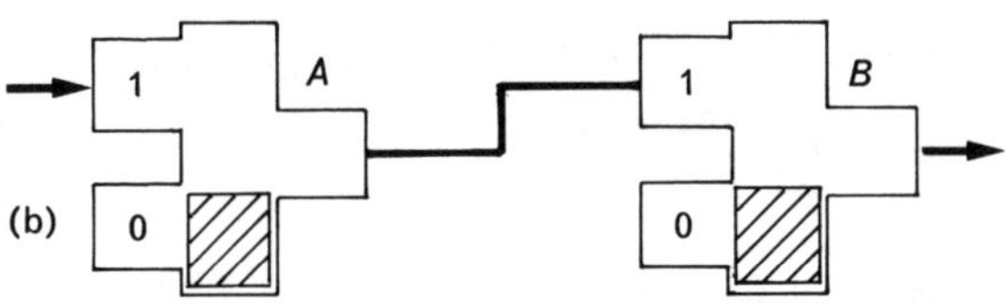

A AND $B = A.B$
A pressure signal to open both valves is required for an output.

(c)

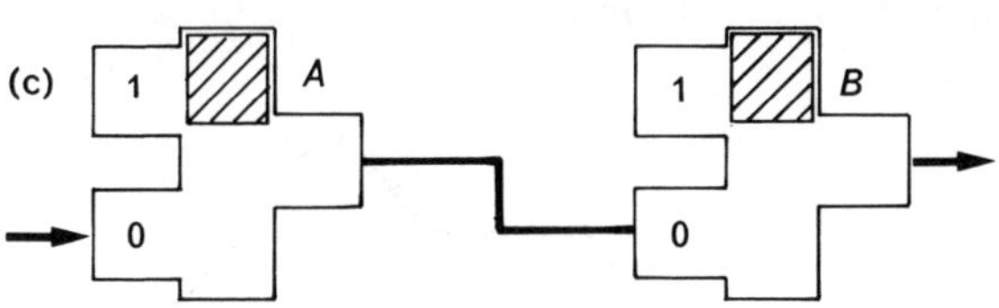

A NOR $B = \overline{A+B} = \overline{A}.\overline{B}$
All input pressure signals must be removed to allow an output. The application of one input signal will interrupt the output.

Fig. 14.3

EXAMPLE 7

Arrange four three-way valves to obtain an output from the function

$$(A+B).\overline{C}.D = 1$$

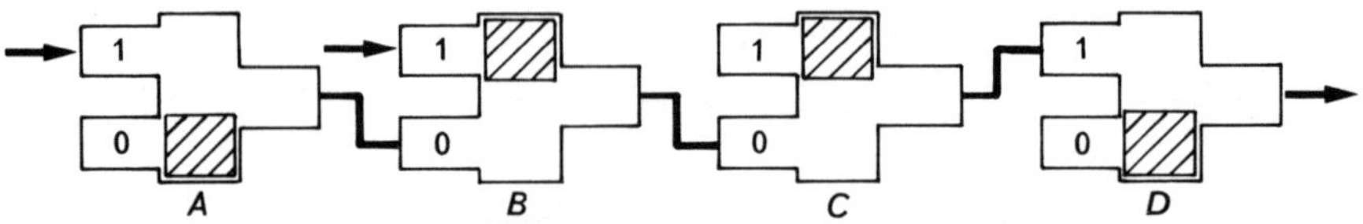

Fig. 14.4

If a pressure is applied to either valve A or valve B then a pressure will be applied to the normally open port of valve C. Thus, provided there is no pressure signal at C, it will remain open and allow the pressure to be applied to valve D to achieve an output. See Fig. 14.4.

Exercise 41

1) If $A.\overline{B}.C = 1$, find the values of A, B and C.

2) If $A.\overline{B}.\overline{C}.D = 1$ find the values of A, B, C and D.

3) Find the values of:

(a) $1.1.1.0.1$ (b) $\overline{0}.1.1.\overline{0}$ (c) $\overline{1}.0.1.\overline{0}.1$

4) Decide which logic element would be required to give a signal to the railway points P_1 and P_2 of Fig. 14.5 so that train T will reach station S.

5) Which logic gate or gates will ensure that train T of Fig. 14.6 arrives at

(a) station S? (b) station S via B?

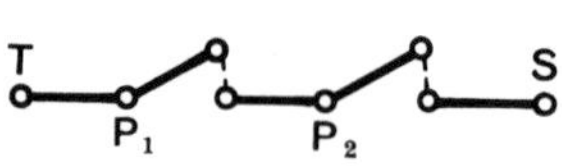

Fig. 14.5

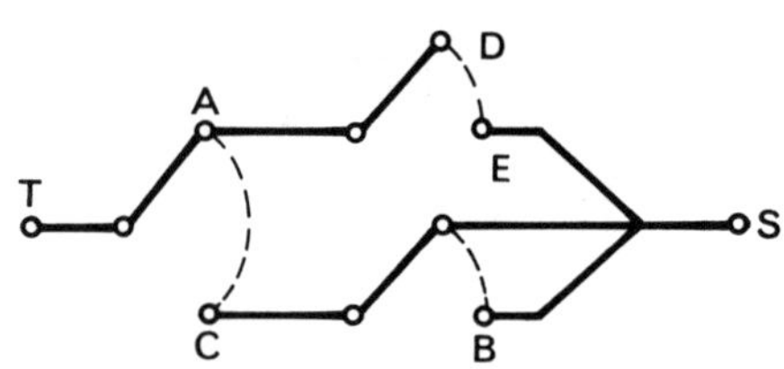

Fig. 14.6

6) Show that by placing the sliding valves in such a way that we obtain an OR, AND, AND, AND and NOT gates, Fig. 14.7 will achieve an output.

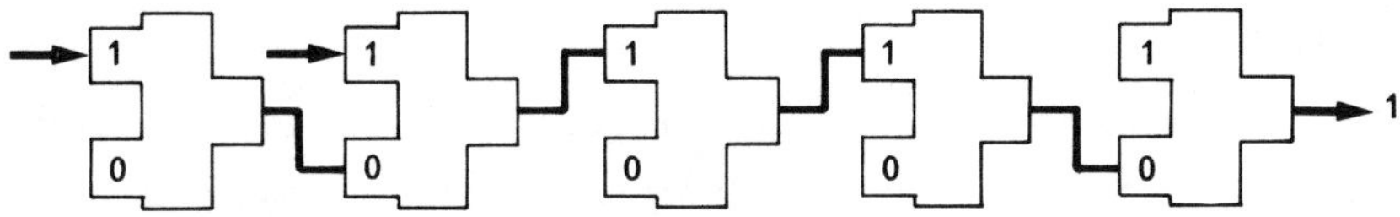

Fig. 14.7

7) Draw logic diagrams which all have inputs A, B and C but have the following output expressions:

(a) $A+BC$ (b) $AB+\overline{C}$ (c) $\overline{A+B}+C$ (d) $\overline{A}+\overline{B}+C$

8) Write down the output expression obtained from each of the following diagrams:

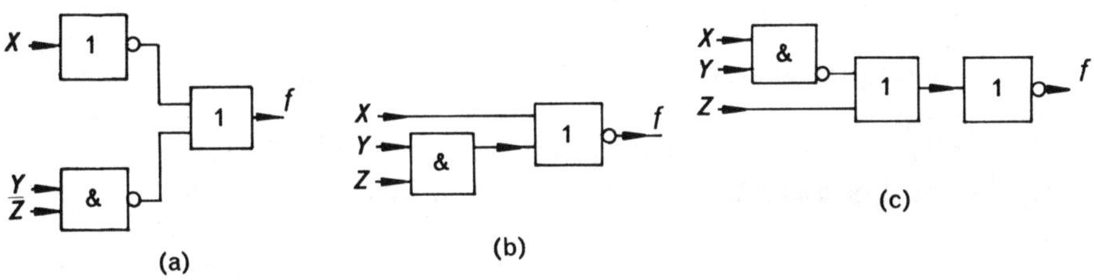

Fig. 14.8

THE COMPUTER

The computer is another tool which is now readily available to technicians. It can be used to store and process information, and to solve problems. To do this it must be supplied with information called **data** and a set of instructions for processing the data called a **program**. The communication is achieved by means of the two fundamental computer components; hardware and software.

Hardware

This is the physical portion of a computer system and is the term used to describe the electrical, mechanical and peripheral devices. One major piece of hardware is the Central Processing Unit (CPU). This is the computer's electrical control centre which performs all the data operations and contains a storage area, called memory, that is used to hold data for a short term during operations. Peripherals are devices which support the CPU. They can be used for long term or permanent storage of data or for communications via terminals, line printers, tapes or discs.

Software

This is the name given to all written procedures and rules that control the computer operations. Programs describe the data and how you wish it to be processed and many languages have been developed for this purpose.

BASIC, 'Beginner's All-purpose Symbolic Instruction Code', is a programming language specifically devised to be readily understood, but, like all computer languages, certain conventions may have to be slightly modified for use with different makes of computers.

BASIC

Example 8 shows a program consisting of nine numbered statements. The computer will execute this program by obeying the instructions in *ascending* order of statement number.

By numbering the statements 10, 20, 30, . . . instead of 1, 2, 3, . . . we are able to insert new instructions, if required, between the existing ones. The VARIABLES, in this example P, Q, R, . . . , can either be a single capital letter or a capital followed by a digit, like A3, K9, Z4, . . .

When typing the program the letters l or I must not be typed for the number 1, nor the letter O for the number 0.

EXAMPLE 8

```
10 LET P = 1.4
20 LET Q = 2
30 LET R = P+Q
40 LET S = P-Q
50 LET T = P*Q
60 LET U = P/Q
70 LET V = P↑Q
80 PRINT R,S,T,U,V
90 END
```

(* is the sign for multiplication)
(/ is the sign for division; ÷ is not allowed)
($P \uparrow 3 = P^3 =$ P*P*P)

Lines 10 to 70 are called **assignment** statements. Line 80 is called a **print** statement, and having instructed the computer to execute the program by typing the word **run**, this instructoin will cause the results
3.4 -0.6 2.8 0.7 1.96 to be produced.

The program in Example 8 is not very useful since it can only be used to calculate the values of R, S, T, U and V for the values of $P = 1.4$ and $Q = 2$.

Example 9 shows how an **input** statement may be used so that the computer may use the program for any values given to P and Q.

After typing RUN, the computer will stop executing the program until the values given to P and Q, the DATA, are typed in.

EXAMPLE 9

```
10 INPUT P,Q
20 LET R = P+Q
30 LET S = P-Q
40 LET T = P*Q
50 LET U = P/Q
60 LET V = P↑Q
70 PRINT R,S,T,U,V
80 END
```

Data `1.4,2` (The word 'Data' is not typed)

As in arithmetic when more than one operation is involved a priority is given to the evaluation of the contents of brackets. Where no brackets exist then expressions involving powers (↑) are calculated before multiplication (*) and division (/). Addition (+) and subtraction (−) are performed last. Brackets should be inserted where necessary since

$$3+4/2 = 5 \quad \text{while} \quad (3+4)/2 = 3.5$$

The computer takes operations, by priority, from left to right.

Exercise 42

1) Correct, where necessary, the notation used for the following BASIC variables:

(a) 6A (b) C (c) A1 (d) K5 (e) t4 (f) V2 (g) n

2) Rewrite the following statements in a suitable form for a BASIC program.

(a) ab (b) $p-q$ (c) $\dfrac{a}{b+c}$ (d) $\dfrac{a}{(b+c)^2}$ (e) x^2+y^2

(f) $\dfrac{4}{3}\pi r^3$ (g) $A\left(1+\dfrac{R}{100}\right)^n$ (h) $\dfrac{n}{2}\{2a+(n-1)d\}$ (i) $\dfrac{a(1-r^n)}{1-r}$

3) Substitute A = 2, B = 3 and C = 4 into the following BASIC expressions to determine their value.

(a) (A+B) ↑ 2 (b) A−(B+C) (c) C/A/B
(d) B−C/A−B (e) A*B/C (f) (A+C)/B−A

4) Write down the computer output of the following:

(a)

```
     100 INPUT X,Y
     200 LET Z = X/Y↑2
     300 PRINT Z
     400 END
Data 27,3
```

(b)

```
     10 INPUT A,B,C
     20 LET Z1 = A*B*C
     30 LET Z2 = A+B+C
     40 LET Z3 = Z1/Z2
     50 PRINT Z3
     60 END
Data 3,4,5
```

5) Write a program which has an input of two numbers, r and h, and uses them to evaluate and print out $v = 3.14r^2h$.

6) Write a program which reads in the three numbers l, b and h and uses them to evaluate $v = lbh$ and then print out the values of l, b, h and v.

REM STATEMENT

At any point in a program we can type in a remark. This may be a title or it may state what is about to be calculated. It is done by typing the letters REM, for REMark, at the beginning of the line and the computer accepts this line simply as a comment and not as an instruction.

GOTO STATEMENT

It is often necessary to alter the order in which instructions are carried out in a program. We can do this by means of a GOTO statement. Thus the line

```
70 GOTO 10
```

would cause the computer to jump from statement 70 to statement 10. However, as the computer executed each instruction from statement 10 it would return to statement 70 and once again be sent to statement 10. To escape from this endless loop we use a conditional statement.

CONDITIONAL STATEMENT

This statement uses the words IF . . . , THEN . . . ; for example

```
60 IF C > 100 THEN 80
```

Thus if some number C becomes greater than 100 the computer will jump to instruction 80, otherwise it will continue to the next statement.

We have IF (condition) THEN (line number)

BASIC uses six symbols to represent relational operators and these are illustrated in the following table:

Mathematical symbol	BASIC symbol	Meaning
$=$	=	A is equal to B
$<$	<	A is less than B
$\leqslant$	< = or = <	A is less than or equal to B.
$>$	>	A is greater than B
$\geqslant$	> = or = >	A is greater than or equal to B
$\neq$	< > or > <	A is not equal to B

FLOW CHARTS

As problems become more complex we can reach a solution by planning a sequence of simple steps which can be shown diagrammatically in a flow chart. In a similar way we can plan a program by first drawing a flow chart.

The following table shows the common convention in 'box' styles for flow charts.

	Start or End box.
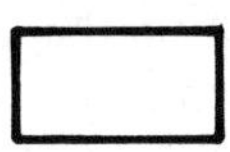	Statement box for instructions.
	Input or Output statement box.
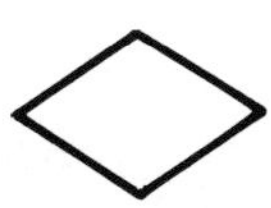	Decision box for questions requiring a decision.

EXAMPLE 10

Write a program to read in 100 numbers, one at a time, and print out the average.

A possible flow chart and program are as follows:

```
10 REM PROGRAM TO CALCULATE THE AVERAGE OF 100 NUMBERS
20 LET S = 0
30 LET C = 0
40 LET C = C+1
50 IF C > 100 THEN 90
60 INPUT X
70 LET S = S+X
80 GOTO 40
90 LET A = S/100
100 PRINT A
110 END
```

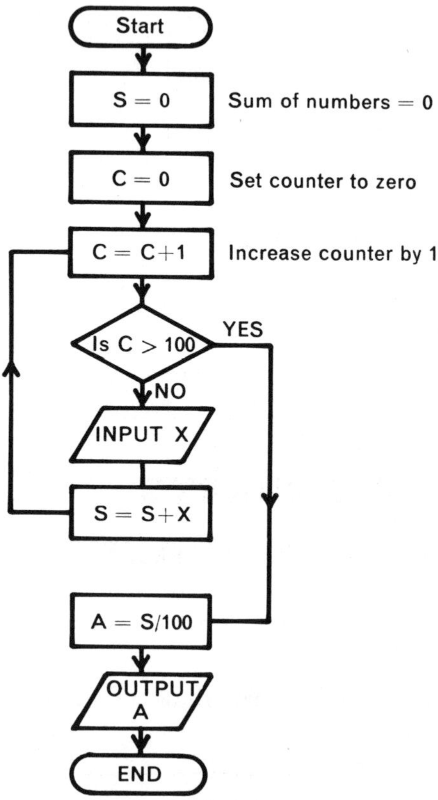

In Example 10, C is used as a counter starting at zero, and on reaching 100 takes us out of the loop as instructed on line 50 of the program. An alternative way of ending a loop is to use 'false data'. For example if all the data is positive we could enter the 101th number in negative form, say −1. Line 50 of the program would then read

```
50 IF C = -1 THEN 90
```

EXAMPLE 11

Items which cost less than £10 have a cash discount of 5 % and those which cost £10 or more have a cash discount of 10 %. Draw a flow chart and write a program to calculate the sale price when the marked price is given as data.

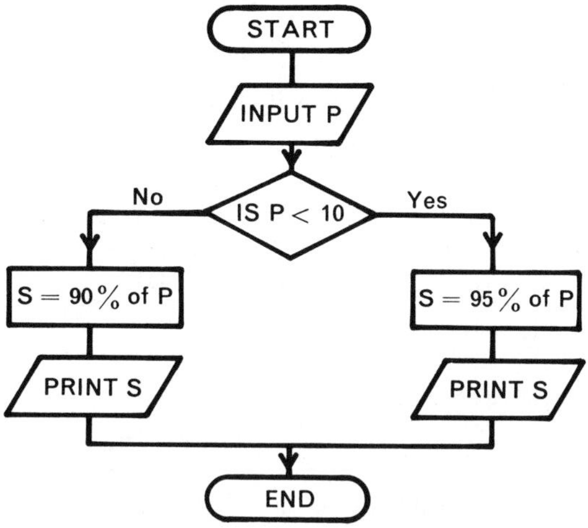


```
10 REM PRICE LESS DISCOUNT
20 INPUT P
30 IF P < 10 THEN 70
40 LET S = P*90/100
50 PRINT "THE SALE PRICE IS £",S
60 GOTO 90
70 LET S = P*95/100
80 PRINT "THE SALE PRICE IS £",S
90 END
```

PRINT STATEMENT

We can output text in BASIC by enclosing it in inverted commas within a PRINT statement as is shown in lines 50 and 80 of Example 11. Thus if $P = £10$ then line 50 would instruct the printer to output

```
THE SALE PRICE IS £9
```

If the statement

```
50 PRINT
```

appears in a program it will cause a 'blank' line to be produced. The introduction of one or more blank lines in this way can improve the appearance of the output.

The use of commas, semicolons and quotation marks in PRINT statements is illustrated in the following example.

EXAMPLE 12

```
10 PRINT 3+2, "EQUALS 3+2"
20 PRINT 3+2; "EQUALS 3+2"
30 PRINT 3+2,,"EQUALS 3+2"
40 END
```

In this example commas are used in line 10 to separate the items while semicolons are used in line 20. In line 30 two commas are used to separate the items. When the program is run, the result is:

```
5          EQUALS 3+2
5EQUALS 3+2
5                    EQUALS 3+2
```

We see that commas separate items into separate 'fields' whereas semicolons cause the data to run together. On the third line the double comma is used to print an empty 'field' between the data items.

A semicolon can also be used to produce two or more program lines to be printed on the same output line. This is illustrated in Example 13.

EXAMPLE 13

```
10 PRINT 3*4/5                 10 PRINT 3*4/5;
20 PRINT 3↑4/5                 20 PRINT 3↑4/5
30 END                         30 END
```

Run

```
 2.4
16.2
```

Run

```
2.4       16.2
```

Exercise 43

1) Write down the output from the following program with input 7,0,1.2 and −999.

```
10 INPUT X
20 PRINT X*X
30 IF X > 0 THEN 10
40 END
```

2) Determine the output, and how it would appear, after running the following program.

```
10 LET X = 2
12 PRINT X;X↑2
14 LET X = X + 2
16 IF X<= 12 THEN 12
18 END
```

3) The formula $A = \frac{1}{2}r^2C$ is used to determine the area of a sector of a circle given the radius r and the angle at the centre C.
Complete the following program to determine the area, given the radius in centimetres and the angle in degrees which *must be converted to radians.*

```
10 LET P = 3.142
20 INPUT R
30 INPUT C
40 LET A =
```

4) Write a program which calculates the income tax payable at the rate of $28\frac{1}{2}\%$ on incomes of £2000 or more. No tax being paid on incomes less than £2000.

5) Write a program which reads in 50 numbers, one at a time, and then prints out each number followed by its square.

6) An engineer receives a bonus of 5% of the sales value of his output up to £5000 and 10% for the amount over £5000. Write a program which calculates his annual bonus.

7) The travelling expenses of a maintenance technician are based on the engine capacity of his car. Up to 1100 cc, 7 p per km; 1101–1500 cc, 9 p per km; over 1500 cc, 12 p per km. Write a program which will read in for each technician his employee number, engine capacity of car, and distance travelled and then output the employee number with his travelling expenses.

SAVING A PROGRAM

When the command NEW is typed the computer will respond with a message like

```
NEW FILE NAME
```

In response to this a program name, different from any other program name, and consisting of up to six characters, should be typed in.

After the program has been typed, the command SAVE causes the computer to copy the program into a storage area.

If the program is then required at a later data the command OLD (or LOAD on some systems) will cause a response like

```
OLD FILE NAME
```

to appear. Having typed in the name of the required program it again becomes available for use.

The command LIST will cause the computer to print a copy of the program on the terminal.

MATHEMATICAL FUNCTIONS

The values of functions such as square roots, sines and cosines are often required by technicians. The computer does not store tables of values of these functions, but calculates them as requested by means of stored programs. The following table lists the usual functions available with BASIC.

Function	Description
ABS(X)	Returns the absolute value of X
ATN(X)	Returns the radian arctangent of the value X
COS(X)	Returns the cosine of the radian value X
COT(X)	Returns the cotangent of the radian value X
EXP(X)	Returns the numeric constant 2.71828 raised to the power of X
INT(X)	Returns the largest integer value of X
LOG(X)	Returns the natural logarithm of the value X
RND(X)	Returns random numbers, (X) is optional
SGN(X)	Returns an indication of the sign (+, −, or 0) of X
SIN(X)	Returns the sine of the radian value X
SQR(X)	Returns the square root of X
TAB(X)	Causes output to be printed at the X column on the output line
TAN(X)	Returns the tangent of the radian value X

Although the *inverse sine* and *inverse cosine* (or arcsine and arc cosine) are not available they can be found from

```
INVERSE SINE = ATN(X/SQR(1 - X*X))

INVERSE COSINE = -ATN(X/SQR(1 - X*X))+1.5708
```

EXAMPLE 14

Write a program to calculate the roots of the quadratic equation $ax^2+bx+c=0$ using the quadratic formula.

```
100 REM THE ROOTS OF A QUADRATIC EQUATION
200 INPUT A,B,C
300 LET X1 = (-B + SQR(B↑2 - 4*A*C))/(2*A)
400 LET X2 = (-B - SQR(B↑2 - 4*A*C))/(2*A)
500 PRINT "THE ROOTS ARE",X1,X2
600 END
```

EXAMPLE 15

Write a program which uses the sine rule to determine a side *a* of triangle ABC given side *b*, angle A and angle B.

Since lower case letters cannot be used in BASIC we must denote the sides of the triangle with capitals. Let $a = \text{A}$, $b = \text{B}$, angle A = A1 and angle B = B1.

Angles measured in degrees must be converted to radian measure, thus

$$A° = A \times \frac{\pi}{180} \text{ radians} = \text{A*3.142/180}$$

and

$$\frac{a}{\text{Sin A}} = \frac{b}{\text{Sin B}} \Rightarrow \text{A} = \text{B*SIN(A1)/SIN(B1)}$$

hence

```
100 REM THE SINE RULE
200 INPUT A1,B1,B
300 LET A2 = A1*3.142/180
400 LET B2 = B1*3.142/180
500 LET A = B*SIN(A2)/SIN(B2)
600 PRINT A
700 END
```

FOR . . . NEXT STATEMENT

Another method of creating a loop in BASIC is by using the combination of FOR and NEXT. See Example 16(a).

The FOR statement acts as a counter for the loop and sets a limit on the number of loops that the program makes. The NEXT statement increments the value in the FOR variable by 1 each time through the loop. This increment, called a STEP, is 1 by default. In Example 16(b) the STEP increment is specified as 2 but this may be any decimal number except 0. If the STEP increment is negative, say −2, then line 10 in Example 16(b) would read

```
10 FOR N = 11 TO 1 STEP -2
```

EXAMPLE 16

(a)

```
NEW

NEW FILE NAME: FOR1

READY

10 FOR N = 1 TO 5 --┐
20 PRINT N, SQR(N) Loop
30 NEXT N -------┘
40 END
RUN
FOR1

1         1
2         1.41421
3         1.73205
4         2
5         2.23607
```

(b)

```
NEW

NEW FILE NAME: FOR2

READY

10 FOR N = 1 TO 10 STEP 2 ┐
20 PRINT N, SQR(N)        |
30 NEXT N---- Loop -------┘
40 END
RUN
FOR2

1         1
3         1.73205
5         2.23607
7         2.64575
9         3
```

Exercise 44

1) Write a program to calculate the length of the hypotenuse in a right-angled triangle.

2) Write a BASIC program to calculate side *b* in triangle ABC given, side *c*, angle B and angle C. (Angles measured in degrees.)

3) Write a BASIC program to calculate side *a* in triangle ABC given, sides *b* and *c* and angle A. (Angles measured in degrees.)

4) Write a BASIC program which outputs the area of any triangle given two sides and the included angle. (Angles measured in degrees.)

5) Write down the output from the following program.

```
10 FOR N = 2 TO 5
20 PRINT N/(N-1);
30 NEXT N
40 END
```

6) Use a FOR loop to print out the decimal equivalents of the fractions

$$\frac{10}{11}, \frac{8}{11}, \frac{6}{11}, \ldots \frac{2}{11}.$$

7) Plot the output from the following program, as coordinates $(x. y)$, on graph paper.

```
     10 LET C = 0.01745
     20 INPUT A,B
     30 FOR X = 0 TO 360 STEP 30
     40 LET Y = A*SIN(C*X) + B
     50 PRINT X,Y
     60 NEXT X
     70 END
Data 3,1
```

8) 100 technicians take an examination. Write a program using a FOR loop to read in for each candidate a number and a mark. The marks are then graded as follows:

Over 79, A; 60–79, B; 45–59, C; less than 45, D.

Print out for each candidate his number, mark and grade.

9) Write a BASIC program which outputs the function values of $f(x) = x^3 - 36x$ for values of x from -6 to 6 in steps of 2. Use the results to draw the function on graph paper.

10) Write a BASIC program which will accept N numbers and calculate their mean and standard deviation.

DIMENSION STATEMENTS

A set of storage locations, within which data can be stored, can be reserved in the computer by stating the DIMension of the storage required. For example

```
10 DIM A(10)
```

would reserve the *eleven* storage locations A(0), A(1), A(2), . . . A(10) since BASIC starts counting from 0. This is acceptable as long as the reserved storage area is greater than what is required.

Similarly the two-dimensional array DIM A(3,3) would reserve the sixteen locations:

```
A(0,0)   A(0,1)   A(0,2)   A(0,3)
        ┌─────────────────────────
A(1,0)  │ A(1,1)   A(1,2)   A(1,3)
A(2,0)  │ A(2,1)   A(2,2)   A(2,3)
A(3,0)  │ A(3,1)   A(3,2)   A(3,3)
```

Lists and arrays like these are generally manipulated using FOR . . . NEXT loops. For example to place 10 numbers into the locations reserved by DIM A(10) we have:

```
10 DIM A(10)
20 FOR G = 1 TO 10
30 INPUT A(G)
40 NEXT G
50 END
```

The first time round the loop, $G = 1$, and the statement INPUT A(G) becomes INPUT A(1), which means the first item of data is stored in location A(1), and so on.

If we inserted a line into the above program which stated

```
45 PRINT A(6)
```

this would print the item stored in that location.

EXAMPLE 17

The maximum error allowed on a product is 0.7 mm. Samples of size 30 are selected and tested. Write a program which will accept the errors and then output each error as a percentage of 0.7 mm.

```
10 REM ERRORS FROM A SAMPLE OF SIZE 30
20 DIM A(30)
30 FOR S = 1 TO 30
40 INPUT A(S)
50 LET P = A(S)/0.7*100
60 PRINT P
70 NEXT S
80 END
```

When writing programs which include loops within loops no overlapping is allowed. This is illustrated in Fig. 14.8.

Correct method of nesting loops

```
10 FOR A =
20 FOR B =
   .
   .
   .
50 NEXT B
60 NEXT A
```

This is improper nesting of loops and is not allowed

```
10 FOR A =
20 FOR B =
   .
   .
   .
50 NEXT A
60 NEXT B
```

Fig. 14.8

EXAMPLE 18

Write a BASIC program which reads in the 15 elements of a (3×5) matrix and then prints out the elements in a rectangular array:

```
10 DIM A(3,5)
20 FOR I = 1 TO 3
30 FOR J = 1 TO 5
40 READ A(I,J)
50 NEXT J
60 NEXT I
70 FOR I = 1 TO 3
80 FOR J = 1 TO 5
90 PRINT A(I,J);
100 NEXT J
110 PRINT
120 NEXT I
130 DATA 4,3,2,1,0
140 DATA 5,-6,-7,8,-9
150 DATA -1,7,2,-3,2
160 END
```

If the data are known when the program is being written we can include them within the program using the READ . . . DATA statement. The two words must be used as a set.

Exercise 45

1) Write a BASIC program to find the mean of 1000 numbers.

2) Write a BASIC program which prints out the largest of 100 numbers.

3) Write a BASIC program which accepts 8 numbers and sorts them into numerical order and then prints them out.

4) Write a program which finds and prints out the sum of the squares of 30 sample numbers.

5) Write a BASIC program which reads in 50 numbers and prints them out in list form.

6) Write a program which reads in the elements of a (3×3) matrix and then prints it out in matrix form.

7) Write a BASIC program which evaluates and prints out the function values of $f(x) = 2x^3-3x^2+36$ for $x\in\{-4, -3, \ldots 3, 4\}$.

PROJECTS

1) The TAB function on page 238 can be used in a sequential system for drilling holes in a numerical controlled machine. Produce a drawing and a manuscript sheet showing the program required to drill a hole at each corner of a small rectangular plate. If facilities are available a punched paper tape could also be produced.

2) Write a computer program (statistical, mechanical, electrical) which produces a result frequently required in your industry.

3) Although records can be held on magnetic disc or drum, punched cards have long been used to store and retrieve information. Make a set of cards to demonstrate their use and construction.

4) Write a simple program in mnemonic code, explaining each step, suitable for a particular microcomputer.

ANSWERS

ANSWERS TO CHAPTER 1

Exercise 1

1) $-5, 19$
2) $39.5, 5, 17$
3) $-4, 8$
4) 3
5) $5.52, -6.52$
6) $2.5, 2, 1$
7) 2

Exercise 2

1) $2x$
2) $7x^6$
3) $12x^2$
4) $30x^4$
5) $1.5t^2$
6) $2\pi R$
7) $\frac{1}{2}x^{-1/2}$
8) $6x^{1/2}$
9) $x^{-1/2}$
10) $2x^{-1/3}$
11) $-2x^{-3}$
12) $-x^{-2}$
13) $-\frac{3}{5}x^{-2}$
14) $-6x^{-4}$
15) $-\frac{1}{2}x^{-3/2}$
16) $-\frac{1}{3}x^{-3/2}$
17) $-\frac{15}{2}x^{-5/2}$
18) $\frac{3}{10}t^{-1/2}$
19) $-0.01h^{-2}$
20) $-5x^{-2}$
21) $8x-3$
22) $9t^2-4t+5$
23) $4u-1$
24) $20x^3-21x^2+6x-2$
25) $35t^4-6t$
26) $\frac{1}{2}x^{-1/2}+\frac{5}{2}x^{3/2}$
27) $-3x^{-2}+1$
28) $\frac{1}{2}x^{-1/2}-\frac{1}{2}x^{-3/2}$
29) $3x^2-\frac{3}{2}x^{-3/2}$
30) $1.3t^{0\cdot3}+0.575t^{-3\cdot3}$
31) $\frac{9}{5}x^2-\frac{4}{7}x-\frac{1}{2}x^{-1/2}$
32) $-0.01x^{-2}$
33) $4.65x^{0\cdot5}-1.44x^{-0\cdot4}$
34) $\frac{3}{2}x^2+5x^{-2}$
35) $-6+14t-6t^2$
36) $-5, 19$
37) $39.5, 5, 17$
38) $-4, 8$
39) 3
40) $5.52, -6.52$
41) $2.5, 2, 1$
42) 2

Exercise 3

1) $4\cos x$
2) $\cos x$
3) $1-4\cos x$
4) $3\cos x$
5) $-3\sin\theta$
6) $1-4\sin t$
7) $2(\theta+\sin\theta)$
8) $-2\sin\theta+3\cos\theta$
9) $9.6t+\frac{1}{4}\sin t$
10) $\dfrac{-10}{v^2}+6\cos v$
11) (a) $2-\sin x$ is always $\geqslant 1$
(b) $\cos t+1$ is always $\geqslant 0$
12) (a) 1 (b) 0 (c) -1
13) 2.67
14) 13.25
15) $32, -32$
16) 8

Exercise 4

1) $5(x+1)^4$
2) $9(3x+2)^2$
3) $4(6+x)^3$
4) $28(4x-1)^6$
5) $-3(3-x)^2$
6) $-9(2-3x)^2$
7) $-4(2-x)^3$
8) $-10(4-2x)^4$
9) $8x(x^2-1)^3$
10) $9x^2(1+x^3)^2$
11) $20x^3(x^4-2)^4$
12) $-10x^4(1-x^5)$
13) $(2t+3)^{-1/2}$
14) $(3\theta-1)^{-1/3}$
15) $-4(r-1)^{-5}$
16) $6(2-3t)^{-3}$
17) $(2h-5)^{-1/2}$
18) $-2(1-4r)^{-1/2}$
19) $6(1-x)^{-2}$
20) $-16(2x+7)^{-3}$
21) $4\cos 4x$
22) $4\cos(4x+3)$
23) $10\pi\cos(2\pi t-0.04)$
24) $-2\sin 2x$
25) $-2\sin(2x+3)$
26) $-4000\pi\sin(100\pi t)$
27) $-500\pi\sin(100\pi t-0.001)$
28) $-15\sin(3t-0.1)$
29) $500\pi\cos(2\pi t)$
30) $100\pi\sin(0.05-2\pi t)$

ANSWERS TO CHAPTER 2

Exercise 5

1) $1+5z+10z^2+10z^3+5z^4+z^5$
2) $p^6+6p^5q+15p^4q^2+20p^3q^3+15p^2q^4+6pq^5+q^6$
3) $x^4-12x^3y+54x^2y^2-108xy^3+81y^4$
4) $32p^5-80p^4q+80p^3q^2-40p^2q^3+10pq^4-q^5$
5) $128x^7+448x^6y+672x^5y^2+560x^4y^3+280x^3y^4+84x^2y^5+14xy^6+y^7$
6) $x^3+3x+\dfrac{3}{x}+\dfrac{1}{x^3}$
7) $1+12x+66x^2+220x^3$
8) $1-28x+364x^2-2912x^3$
9) $p^{16}+16p^{15}q+120p^{14}q^2+560p^{13}q^3$
10) $1+30y+405y^2+3240y^3$
11) $x^{18}-27x^{16}y+324x^{14}y^2-2268x^{12}y^3$
12) $x^{22}+11x^{18}+55x^{14}+165x^{10}$
14) 5.6% too large
15) 10.2% too small
16) 2.9% decrease
17) 20.5% increase

Exercise 6

1) $1/9\left(1-\frac{2x}{3}\right)^{-2}$ **2)** $4\left(1+\frac{x}{16}\right)^{1/2}$

3) $\frac{1}{3}\left(1-\frac{x}{9}\right)^{-1/2}$ **4)** $4(1+0.3x)^{2/3}$

5) $1+6x+24x^2+80x^4$; $|x|<\frac{1}{2}$

6) $\frac{1}{2}(1-1/8x+3/128x^2-5/1024x^3)$; $|x|<4$

7) $1+x+\frac{3}{2}x^2+\frac{5}{2}x^3$; $|x|<\frac{1}{2}$

8) $3(1-0.1x-0.005x^2-0.0005x^3)$; $|x|<5$

9) 1.024695

ANSWERS TO CHAPTER 3

Exercise 7

1) a^{11} **2)** z^{11} **3)** y^{12}
4) 2^8 **5)** 3^8 **6)** $\frac{3}{32}a^6$
7) a^3 **8)** m^7 **9)** 2^4
10) x^{15} **11)** a^4 **12)** q^8
13) m **14)** l^2 **15)** L^2
16) x^{12} **17)** a^{15} **18)** $9x^8$
19) 2^6 **20)** 10^6 **21)** a^3b^6
22) $a^4b^8c^{12}$ **23)** $2^5x^{10}y^{15}z^5$
24) $\frac{3^5m^{10}}{4^5n^{15}}$ **25)** $\frac{1}{10}, \frac{1}{4}, \frac{1}{81}, \frac{1}{25}$
26) 32, 625, 2, 3 **27)** $3^6, 3^{15}, 3^{12}, 3^{17}$
28) $a^{1/5}, a^{2/3}, a^{4/7}, a^3$ **29)** 64, 1 000 000,4,8
30) 4, 8, $\frac{1}{3}$, $\frac{1}{3}$ **31)** 10
32) $\frac{5}{6}$ **33)** 1, $\frac{1}{5}$, 100 000
34) $\frac{1}{3x^2}, \frac{1}{3x^2}, \frac{1}{5a^4}$ **35)** $32a^5$

Exercise 8

1) 3 **2)** 3 **3)** 4 **4)** 3 **5)** 9
6) 64 **7)** 100 **8)** 1 **9)** 2 **10)** 3

Exercise 9

1) 2.099 4 **2)** 0.392 77
3) 0.739 28 **4)** 3.453 7
5) 0.330 72 **6)** 0.099 276
7) 0.748 79 **8)** 1.849 2
9) 4.530 1 **10)** 0.071 529
11) 2.084 5 **12)** 368.48
13) 0.795 42 **14)** 0.000 179 16
15) 0.266 99 **16)** 3.594 4
17) 3.059 7 **18)** 0.093 370
19) 3.559 5 **20)** 6.664 6
21) 0.523 33 **22)** 201.18
23) 5.065 3 **24)** 1.199 6
25) 1.773 8 **26)** −7.379 3 **27)** −0.322 47
28) 4.627 6 **29)** 2.354 4 **30)** −2.278 5
31) −36.536 6 **32)** 22.255 9

Exercise 10

1) (a) 5.741 (b) 1.957 (c) 3.003
(d) 2.095 (e) −0.711 (f) −4.806
2) (a) 3.584 (b) 6.178 (c) −0.0456
(d) −3.267 (e) 0.728 (f) −2.303

Exercise 11

1) (a) 2.212 (b) 1.795 (c) 3.373
(d) 5.698 (e) −0.525 (f) −4.942
2) 0.00319 **3)** 0.0146
4) (a) 1.35 (b) 12.18 (c) 90.02
(d) 0.67 (e) 0.37 (f) 0.03
5) $y = 1.82$, $x = 0.84$
6) $x = -1.39$, $y = -0.5$
7) (a) $T = 631$ (b) $s = 1.12$
8) 4.34 mA per second
9) (a) 0.18 seconds
(b) 1200 V per second
10) 10 years

ANSWERS TO CHAPTER 4

Exercise 12

1) $a = 94$, $b = 0.35$ **2)** $E = 1.5$, $r = 0.6$
3) $a = 3$, $n = 0.5$
4) n = 4, for $V = 80$ read $V = 70$
5) $t = 0.3x^{1.5}$
6) $k = 100$, $n = -1.2$
7) $a = 245$, $b = 33$ **8)** $\theta = 0.5$, $k = 5$
9) $I = 0.02$, $T = 0.2$

ANSWERS TO CHAPTER 5

Exercise 13

1) ±0.79, ±2.52 **2)** 30°, 90°, 150°
3) −1/8, 8 **4)** 1/9, 9 **5)** ±3, ±4
6) ±4, ±5 **7)** 2.0794 **8)** 8.27, 4.23
9) 18 **10)** 99×63 or 90×72
12) 2.5 **13)** −1/5, $\frac{1}{2}$ **14)** 2 m

Exercise 14

1) $x = 0.333$, $y = -1.667$;
$x = 2$, $y = 5$
2) $x = -1$, $y = 6$;
$x = 0.375$, $y = -0.875$
3) $x = 1.175$, $y = 2.175$;
$x = -0.425$, $y = 0.575$
4) $x = -0.969$, $y = 0.956$;
$x = 0.579$, $y = -1.256$
5) $x = 0.211$, $y = -1.422$:
$x = -1.354$, $y = 1.708$
6) $x = 0.167$, $y = 0.805$;
$x = 0.833$, $y = 2.278$
7) $x = -4$, $y = 2$;
$x = 8$, $y = -2$
8) $x = -1$, $y = 1$;
$x = 4$, $y = 6$
9) $x = 0$, $y = 5$;
$x = 3$, $y = 11$
10) $x = 10.8$, $y = 3.89$;
$x = 2.23$, $y = -0.39$
11) $x = 1$, $y = 3$;
$x = -0.2$, $y = -3$
12) $x = 2.39$, $y = 6.91$;
$x = 0.26$, $y = 0.54$

Exercise 15

1) $\{(0, 0, 0)\}$ **2)** $\{(-1, -2, -3)\}$
3) $\{(-2, -1, 1)\}$ **4)** $\{(1, 1, 1)\}$
5) $\{(3, -2, -\frac{1}{2})\}$ **6)** $\{(1, -2, 0)\}$
7) $\{(1, 0, -3)\}$ **8)** $\{(12, -12, 12)\}$
9) $\{(3, 5, 2)\}$ **10)** $\{(1, -1, 2)\}$

Exercise 16

1) 0.3168 **2)** 1.456 **3)** 2.28
4) 1.895 **5)** 2.130

ANSWERS TO CHAPTER 6

Exercise 17

1) 38 **2)** $\frac{41}{12}$ **3)** $-9\sqrt{2}$
4) 5, 9, 11 **5)** $7\frac{3}{4}$, $12\frac{1}{2}$, $17\frac{1}{4}$
6) 13, 3, −2 **7)** $5\frac{2}{3}$, $9\frac{1}{3}$, $16\frac{2}{3}$
8) 21 **9)** 192 **10)** 2
11) $\frac{9}{4}, \frac{3}{2}, 1$ **12)** $\frac{32}{243}$ **13)** £1305
14) $3\frac{3}{4}$, $7\frac{1}{2}$, 15, 30, 60, 120

Exercise 18

1) 531 **2)** 1575 **3)** 16
4) 1683 **5)** 50 **6)** 63
7) 21 **8)** £258 750 **10)** 63.7, 240.1, 10

Exercise 19

1) 256, 508 **2)** $1/3^8$, 13.5
3) 13 122, 19 682 **4)** $1/4^9$, 1.3333
5) 4^6, 5461 **6)** $\frac{2}{4^{18}}$, 10.667
7) $\frac{5}{4}$ **8)** $\frac{3}{2}$ **9)** $\frac{16}{3}$
10) None **11)** 6/5 **12)** None
13) 1023, 7 **14)** £$(2^{101}-1)$
16) 273.14, 800 **17)** 53 **18)** 70.4
19) £15.03×10^6, £20×10^6

Exercise 20

1) (a) $\frac{1}{2}, \frac{3}{7}, \frac{13}{30}, \frac{29}{67}$
(b) $1, \frac{2}{3}, \frac{9}{13}, \frac{29}{42}, \frac{67}{97}, \frac{230}{333}, \frac{987}{1429}, \frac{1217}{1762}, \frac{5855}{8477}$
(c) $2, \frac{5}{2}, \frac{7}{3}, \frac{19}{8}, \frac{64}{27}, \frac{147}{62}$
2) 0.0017 mm **4)** $\frac{85}{115}$ **5)** 34 on 71
6) 34 on 37, 1′

ANSWERS TO CHAPTER 7

Exercise 21

1) $\{53.13, 306.87\}$ **2)** $\{14.48, 165.52\}$
3) $\{145, 325\}$ **4)** $\{225, 315\}$
5) $\{150, 330\}$ **6)** $\{210, 330\}$
7) $\{17.5, 162.5, 197.5, 342.5\}$
8) $\{78.5, 101.5, 258.5, 281.5\}$
9) $\{50.2, 129.8, 230.2, 309.8\}$

Exercise 22

1) $\{7\frac{1}{2}, 37\frac{1}{2}, 97\frac{1}{2}, 127\frac{1}{2}, 187\frac{1}{2}, 217\frac{1}{2}, 277\frac{1}{2}, 307\frac{1}{2}\}$
2) $\{40, 80, 160, 200, 280, 320\}$
3) $\{15, 75, 135, 195, 255, 315\}$

4) {33, 87} **5)** {58, 152} **6)** {45.7, 74.3}
7) {76, 256}

Exercise 23

1) {70.53, 180, 289.47}
2) {76, 135, 256, 315}
3) {66.4, 180, 293.6}
4) {30, 41.8, 138.2, 150}
5) {120, 131.8, 228.2, 240}
6) {30, 150, 217, 323} **7)** {53.1}
8) {167.3, 347.3} **9)** {333.4}
10) {180, 270}

Exercise 24

1) 10.2 **2)** 56.64°, 7.05°
3) 7.18° **4)** 58.75°, 4.89°
5) 5.2, 1.97 **6)** 8.68, 91.03°

Exercise 25

1) 2, −1
2) (a) $\begin{pmatrix}4\\-5\\2\end{pmatrix}$ (b) $\begin{pmatrix}4\\-7\\12\end{pmatrix}$ (c) $\begin{pmatrix}10\\-6\\16\end{pmatrix}$ (d) $\begin{pmatrix}-4\\-10\\-14\end{pmatrix}$
3) (a) $\sqrt{3}$ (b) $\sqrt{26}$ (c) 1
4) (a) 1, $\frac{1}{2}$, −3 (b) −4, 5, 1
5) 23.18 **6)** 29.24, 074° **7)** 70.7 N
8) 625 N **9)** $\mathbf{i}-2\mathbf{j}$ **10)** 120°
11) (6, −3, 1)
12) (4, −1, 1), (6, 7, 3), (0, 9, 11)
13) (a) (5, $3\frac{1}{2}$, 1) (b) $\left(\frac{10}{3}, \frac{8}{3}, -\frac{2}{3}\right)$
14) 3:1
16) (a) 3.6 (b) 1.4 against (c) 1.4
18) 109.5° **19)** $\frac{t}{5}(2\mathbf{i}+4\mathbf{j}+\sqrt{5}\mathbf{k})$
21) 17 units **22)** −12, 80, 8, 76
23) 76.23°, 10.2 **24)** $\frac{1}{3}\begin{pmatrix}-1\\2\\2\end{pmatrix}$

ANSWERS TO CHAPTER 8

Exercise 26

1) (a) 50.42 mm and 0.225 mm
(b) 50.42 and 0.223 (c) 22
2) (a) 0.905 0 (b) 0.011 9
(c) 0.003 3
3) (a) 8 (b) 4.472 (c) 8
(d) 3.162 (e) $\mu = \bar{x} = 8$;
$s = \frac{\sigma}{\sqrt{n}} = 3.162$
4) (a) 635 g and 0.233 g (b) 635 g and 0.232 g
5) (a) 19.048 mm (b) 0.048 9 mm
6) (a) 172.7 cm (b) 5.0 cm
7) (a) (i) $\frac{2}{3}$ (ii) $\frac{2}{5}$ (iii) $\frac{2}{10}$
(b) (i) 25.3±1.31 (ii) 25.3±0.784
(iii) 25.3±0.392
8) (a) 14.1±2.00%
(b) 14.1±2.87%
9) (a) 53.6±0.60% mm
(b) 53.6±0.87% mm
10) (a) 53.6±0.50 mm
(b) 53.6±0.65 mm

Exercise 27

1) 40.0138, 40.0088, 39.9912, 39.9862
5) 1000, £987.50 **6)** 200

ANSWERS TO CHAPTER 9

Exercise 28

1) $\frac{x^3}{3}+c$ **2)** $\frac{x^9}{9}+c$
3) $\frac{2x^{3/2}}{3}+c$ **4)** $-\frac{1}{x}+c$
5) $-\frac{1}{3x^3}+c$ **6)** $\frac{3x^5}{5}+c$
7) $2x^{1/2}+c$ **8)** $\frac{5x^9}{9}+c$
9) $\frac{x^3}{3}+\frac{x^2}{2}+3x+c$ **10)** $\frac{x^4}{2}-\frac{7x^2}{2}-4x+c$
11) $\frac{x^3}{3}-\frac{5x^2}{2}+2x^{1/2}-\frac{2}{x}+c$
12) $-\frac{4}{x^2}+\frac{2}{x}+\frac{2x^{3/2}}{3}+c$
13) $\frac{x^3}{3}-\frac{3x^2}{2}+2x+c$ **14)** $\frac{x^3}{3}+3x^2+9x+c$
15) $\frac{4x^3}{3}-14x^2+49x+c$

Exercise 29

1) $y=\frac{x^2}{2}+1$ 2) 46.7
3) $y = 10+3x-x^2$, 10
4) $y=\frac{2x^3}{3}+\frac{3x^2}{2}+2x-\frac{1}{6}$
5) $p=\frac{t^3}{3}-3t^2+9t-\frac{17}{3}$
7) $y = x^2-2x+4$ 8) $y=\frac{x^3}{3}+1$
9) 3.33

Exercise 30

1) 2.33 2) 8 3) 8.67
4) 4 5) 3.167 6) 1.89
7) 0.667 8) 0.667

Exercise 31

1) 14.7 2) 2.67 3) −0.167
4) 4.5 5) +0.25, −0.25
7) 2.67 8) 2.19 10) 0.866
11) 57.2 12) 16

Exercise 32

1) 57.4 2) 402 3) 0.0761
4) 171 5) 16.8 6) 0.105
7) (a) 262 000 mm^3 (b) 58 700 mm^3
8) 14.6 litres

ANSWERS TO CHAPTER 10

Exercise 33

1) $\bar{x} = 38.7$ mm, $\bar{y} = 41.8$ mm
2) $\bar{x} = 68.3$ mm, $\bar{y} = 73.8$ mm
3) $\bar{x} = 19.1$ mm, $\bar{y} = 28.3$ mm
4) $\bar{x} = 50.0$ mm, $\bar{y} = 44.6$ mm
5) $\bar{y} = 42.3$ mm
6) $\bar{x} = 86.7$ mm, $\bar{y} = 60.7$ mm
7) $\bar{y} = 29.7$ mm 8) $\bar{y} = 10.6$ mm

ANSWERS TO CHAPTER 11

Exercise 34

1) $I_{XX} = 856$ cm^4, $I_{YY} = 1184$ cm^4
2) $I_{XX} = 1410$ cm^4, $I_{YY} = 354$ cm^4
3) $I_{XX} = 296$ cm^4, $I_{YY} = 133$ cm^4
4) $I_{XX} = 521$ cm^4, $I_{YY} = 1100$ cm^4
5) $I_{XX} = 492$ cm^4, $I_{YY} = 172$ cm^4
6) $I_{XX} = 136$ cm^4, $I_{YY} = 60$ cm^4

Exercise 35

1) $I_{XX} = 375$ cm^4, $I_{YY} = 266$ cm^4
2) $I_{XX} = 131$ cm^4, $I_{YY} = 227$ cm^4
3) $I_{XX} = 273$ cm^4, $I_{YY} = 216$ cm^4
4) $I_{XX} = I_{YY} = 168$ cm^4
5) 1410 cm^4, 296 cm^4, 521 cm^4, 492 cm^4
6) (a) 258 cm^4 (b) 517 cm^4
(c) 2520 cm^4
7) (a) 1600 cm^4 (b) 1820 cm^4
(c) 909 cm^4

ANSWERS TO CHAPTER 12

Exercise 36

1) 0.0133 kg m^2 2) 104 mm
3) 21.4 kg m^2
4) (a) 8390 kg m^2 (b) 32 300 kg m^2
5) (a) 288 kg m^2 (b) 1050 kg m^2
6) 17 700 kg m^2 7) 5.83 kg m^2

ANSWERS TO CHAPTER 13

Exercise 37

1) $18x$, 54 2) −187, 254
3) −20 4) 143.2
5) 15, 0, −135 6) −89
7) −0.115 8) 51583 9) 0, 8.66

Exercise 38

1) 42 m/s 2) 6 m/s^2
3) (a) 6 m/s (b) 2.41 or −0.41 s
(c) 6 m/s^2 (d) 1 s
4) −0.074 m/s, 0.074 m/s^2
5) 10 m/s, 30 m/s 6) 3.46 m/s
7) (a) 4 rad/s (b) 36 rad/s^2
(c) 0 s or 1 s
8) (a) −2.97 rad/s (b) 0.280 s
(c) −8.98 rad/s^2 (d) 1.57 s
9) 62.5 kJ

Exercise 39

1) (a) 11 (max), −16 (min)
   (b) 4 (max), 0 (min)
   (c) 0 (min), 32 (max)
2) (a) 54 (b) $x = 2.5$ (c) $x = -2$
3) (3, −15), (−1, 17)
4) (a) −2 (b) 1 (c) 9
5) (a) 12 (b) 12.48
6) 15 mm 7) 10 m 8) 4
9) 108 000 mm^3
10) radius = height = 4.57 m
11) 405 mm
12) diameter = 28.9 mm, height 14.4 mm
13) 5.76 m

ANSWERS TO CHAPTER 14

Exercise 40

1) (a) {3, 4, 5, 6, 7, 8, 9, 11} (b) {4, 6, 8}
   (c) C (d) {3, 5, 7}
   (e) {2, 4, 6, 8, 9, 10, 11}
   (f) {2, 4, 6, 8, 9, 10, 11} (g) {2, 10}
   (h) {2, 4, 6, 8, 9, 10, 11}
2) (a) 0 (b) 1 (c) A (d) Y (e) X
4) (a) 0 (b) $X+Y(W+Z)$

Exercise 41

1) 1, 0, 1 2) 1, 0, 0, 1
3) (a) 0 (b) 1 (c) 0 4) AND
5) (a) E AND NOT A ($E.\overline{A}$)
   (b) C AND NOT A AND B AND NOT E ($C.\overline{A}.B.\overline{E}$)
7)

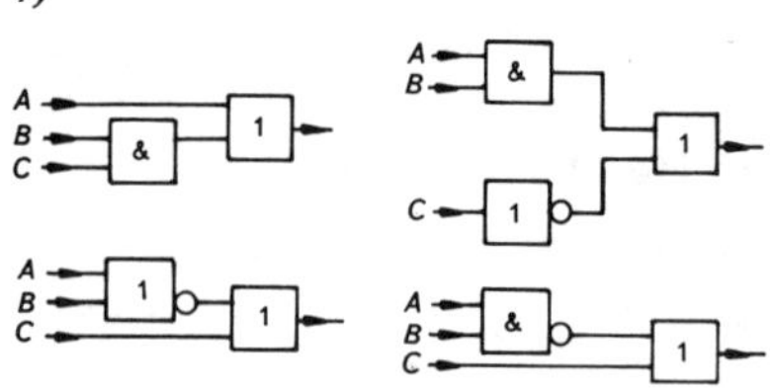

8) (a) $\overline{X}+\overline{YZ}$ (b) $\overline{X+YZ}$ (c) $XY\overline{Z}$

Exercise 42

1) (a) A6 (e) T4 (g) N
2) (a) A*B (b) P−Q (c) A/(B+C)
   (d) A/(B+C)↑2 (e) X*X + Y*Y
   (f) 4*3.142*R*R/3
   (g) A*(1+R/100)↑N
   (h) N*(2*A+(N−1)*D)/2
   (i) A*(1−R↑N)/(1−R)
3) (a) 25 (b) −5 (c) $\frac{2}{3}$ (d) 1 (e) $1\frac{1}{2}$ (f) 6
4) (a) 3 (b) 5
5)
```
10 INPUT R,H
20 LET V = 3.142*R*R*H
30 PRINT V
40 END
```
6)
```
10 INPUT L,B,H
20 LET V = L*B*H
30 PRINT L,B,H,V
40 END
```

Exercise 43

1) 49, 0, 1.44
2)
```
 2   4
 4  16
 6  36
 8  64
10 100
12 144
```
3)
```
40 LET A + 0.5*R*R*C*P/180
50 PRINT A
60 END
```
4)
```
10 REM INCOME TAX
20 INPUT I
30 IF I<2000 THEN 70
40 LET T = (I-2000)*28.5/100
50 PRINT T
60 GOTO 80
70 PRINT "NO TAX"
80 END
```
5)
```
10 LET N = 0
20 LET N = N+1
30 IF N>50 THEN 80
40 INPUT X
50 LET S = X*X
60 PRINT X,S
70 GOTO 20
80 END
```
6)
```
10 REM BONUS
20 INPUT S
30 IF S<5000 THEN 70
40 LET B2=((S-5000)*5/100) + 500
50 PRINT B2
60 GOTO 90
70 LET B1 = S*10/100
80 PRINT B1
90 END
```

7)
```
10 REM EXPENSES E, NUMBER N
20 REM CAPACITY C, MILES M
30 INPUT N,C,M
40 IF C>1500 THEN 100
50 IF C>1100 THEN 80
60 LET E = M*7
70 GOTO 110
80 LET E = M*9
90 GOTO 110
100 LET E = M*12
110 PRINT N,E
120 END
```

8)
```
10 REM EXAM RESULTS
20 FOR C = 1 TO 100
30 INPUT N,M
40 IF M>79 THEN 130
50 IF M>59 THEN 110
60 IF M>44 THEN 90
70 PRINT N,M,"D"
80 GOTO 140
90 PRINT N,M,"C"
100 GOTO 140
110 PRINT N,M,"B"
120 GOTO 140
130 PRINT N,M,"A"
140 NEXT C
150 END
```

9)
```
10 REM FUNCTION VALUES
20 FOR X = -6 TO 6 STEP 2
30 LET F = X↑3 -36*X
40 PRINT F
50 NEXT X
60 END
```

10)
```
10 REM MEAN AND ST DEV
20 INPUT N
30 LET S1 = 0
40 LET S2 = 0
50 FOR I = 1 TO N
60 INPUT X
70 LET S1 = S1 + X
80 LET S2 = S2 * X↑2
90 NEXT I
100 LET M = S1/N
110 LET V = (S2-(S1)↑2/N)/N
120 LET D = SQR(V)
130 PRINT "MEAN=",V;"ST DEV=",D
140 END
```

Exercise 44

1)
```
10 REM HYPOTENUSE A
20 INPUT B,C
30 LET X = (B*B)+(C*C)
40 LET A = SQR(X)
50 PRINT "HYPOTENUSE X =",A
60 END
```

2)
```
10 REM TRIANGLE SIDE
20 REM DEGREES TO RADIANS X P
30 INPUT C,B1,C1
40 LET P = 0.01745
50 LET B2 = B1*P
60 LET C2 = C1*P
70 LET B = C*SIN(B2)/SIN(C2)
80 PRINT B
90 END
```

3)
```
10 REM COSINE RULE
20 INPUT B,C,A1
30 LET P = 0.01745
40 LET A2 = A1*P
50 LET A = B↑2+C↑2-2*B*C*COS(A2)
60 LET A3= SQR(A)
70 PRINT A3
80 END
```

4)
```
10 REM AREA OF TRIANGLE
20 INPUT B,C,A1
30 LET P = 0.01745
40 LET A2 = A1*P
50 LET X = 0.5*B*C*SIN(A2)
60 PRINT X
70 END
```

5) 2   1.5   1.333333   1.25

6)
```
10 REM FRACTIONS
20 FOR N = 10 TO 1 STEP -2
30 PRINT N/11
40 NEXT N
50 END
```

Exercise 45

1)
```
10 REM MEAN
20 DIM A(1000)
30 LET S = 0
40 FOR X = 1 TO 1000
50 INPUT A(X)
60 LET S = S+A(X)
70 NEXT X
80 LET M = S/1000
90 PRINT M
100 END
```

2)
```
10 REM LARGEST NUMBER
20 DIM A(100)
30 LET L = 0
40 FOR N = 1 TO 100
50 INPUT A(N)
```

```
 60 IF A(N)<L THEN 80
 70 LET L = A(N)
 80 NEXT N
 90 PRINT L
100 END
```

3)
```
 10 DIM A(8)
 20 FOR I = 1 TO 8
 30 INPUT A(I)
 40 NEXT I
 50 F = 0
 60 FOR I = 1 TO 7
 70 IF A(I)<=A(I+1) THEN 140
 80 T = A(I)
 90 A(I) = A(I + 1)
100 A(I+1) = T
110 F = 1
120 NEXT I
130 IF F = 1 THEN 70
140 FOR I = 1 TO 8
150 PRINT A(I)
160 NEXT I
170 END
```

4)
```
 10 REM SAMPLE SUM OF SQUARES
 20 DIM S(30)
 30 LET X = 0
 40 FOR C = 1 TO 30
 50 INPUT S(C)
 60 LET X = X+S(C)*S(C)
 70 NEXT C
 80 PRINT X
 90 END
```

5)
```
 10 DIM L(50)
 20 FOR X = 1 TO 50
 30 INPUT L(X)
 40 NEXT X
 50 FOR X = 1 TO 50
 60 PRINT L(X)
 70 NEXT X
 80 END
```

6) See Example 18

7)
```
 10 REM FUNCTION VALUES
 20 DIM F(9)
 30 LET X = 1
 40 FOR C = -4 TO 4
 50 LET F(X) = 2*C↑3-3*C↑2+36
 60 LET X = X+1
 70 NEXT C
 80 FOR X = 1 TO 9
 90 PRINT F(X)
100 NEXT X
110 END
```

INDEX